建筑与都市系列丛书 | 世界建筑
Architecture and Urbanism Series | World Architecture

文筑国际 编译
Edited by CA-GROUP

Contemporary Sri Lanka on Geoffrey Bawa's 100th

斯里兰卡：
当代与巴瓦100周年

中国建筑工业出版社

图书在版编目（CIP）数据

斯里兰卡：当代与巴瓦100周年 = Contemporary Sri Lanka on Geoffrey Bawa's 100th. 汉英对照 / 文筑国际 CA-GROUP 编译. -- 北京：中国建筑工业出版社，2021.1

（建筑与都市系列丛书. 世界建筑）

ISBN 978-7-112-25721-8

Ⅰ. ①斯... Ⅱ. ①文... Ⅲ. ①建筑艺术 - 介绍 - 斯里兰卡 - 汉、英 Ⅳ. ① TU-863.58

中国版本图书馆 CIP 数据核字 (2020) 第 247369 号

责任编辑：毕凤鸣 刘文昕
版式设计：文筑国际
责任校对：王烨

建筑与都市系列丛书｜世界建筑
Architecture and Urbanism Series ｜ World Architecture
斯里兰卡：当代与巴瓦 100 周年
Contemporary Sri Lanka on Geoffrey Bawa's 100th
文筑国际　编译
Edited by CA-GROUP
*
中国建筑工业出版社出版、发行（北京海淀三里河路 9 号）
各地新华书店、建筑书店经销
北京雅昌艺术印刷有限公司 制版、印刷
*
开本：787 毫米 ×1092 毫米　1/16　印张：17½　字数：550 千字
2024 年 7 月第一版　2024 年 7 月第一次印刷
定价：148.00 元
ISBN 978-7-112-25721-8
　　　　（36506）

版权所有　翻印必究
如有内容及印装质量问题，请联系本社读者服务中心退换
电话：（010）58337283　QQ：2885381756
（地址：北京海淀三里河路 9 号中国建筑工业出版社 604 室　邮政编码 100037）

a+u

建筑与都市系列丛书学术委员会
Academic Board Members of Architecture and Urbanism Series

委员会顾问 Advisors
郑时龄 ZHENG Shiling   崔 愷 CUI Kai   孙继伟 SUN Jiwei

委员会主任 Director of the Academic Board
李翔宁 LI Xiangning

委员会成员 Academic Board
曹嘉明 CAO Jiaming   张永和 CHANG Yungho   方 海 FANG Hai
韩林飞 HAN Linfei   刘克成 LIU Kecheng   马岩松 MA Yansong
裴 钊 PEI Zhao   阮 昕 RUAN Xing   王 飞 WANG Fei
王 澍 WANG Shu   赵 扬 ZHAO Yang   朱 锫 ZHU Pei

*委员会成员按汉语拼音排序（左起）
Academic board members are ranked in pinyin order from left.

建筑与都市系列丛书
Architecture and Urbanism Series

总策划 Production
国际建筑联盟 IAM   文筑国际 CA-GROUP

出品人 Publisher
马卫东 MA Weidong

总策划人/总监制 Executive Producer
马卫东 MA Weidong

内容担当 Editor in Charge
吴瑞香 WU Ruixiang

助理 Assistants
卢亭羽 LU Tingyu   杨 文 YANG Wen   杨紫薇 YANG Ziwei

特约审校 Proofreaders
黄一凡 HUANG Yifan   曾文菁 ZENG Wenjing   寇宗捷 KOU Zongjie

翻译 Translators
中译英 English Translation from Chinese:
樊梦莹 FAN Mengying (pp.6-7) 杨紫薇 YANG Ziwei (p274)
英译中 Chinese Translation from English:
樊梦莹 FAN Mengying (pp.14-269)
日译中 Chinese Translation from Japanese:
吴瑞香 WU Ruixiang (p13) 万轶群 WAN Yiqun (p277)

书籍设计 Book Design
文筑国际 CA-GROUP

中日邦交正常化50周年纪念项目
The 50th Anniversary of the Normalization of
China-Japan Diplomatic Relations

本系列丛书部分内容选自A+U第590号（2018年09月号）特辑
原版书名：
スリランカの現在——ジェフリー・バワ100周年
Contemporary Sri Lanka on Geoffrey Bawa's 100th
著作权归属A+U Publishing Co., Ltd. 2019

A+U Publishing Co., Ltd.
发行人／主编：吉田信之
客座主编：钱纳·达斯瓦特
副主编：横山圭　Sylvia Chen
编辑：Grace Hong　小野寺谅朔　佐藤绫子
海外协助：侯蕾

Part of this series is selected from the original a+u No. 590 (18:09),
the original title is:
スリランカの現在——ジェフリー・バワ100周年
Contemporary Sri Lanka on Geoffrey Bawa's 100th
The copyright of this part is owned by A+U Publishing Co., Ltd. 2019.

A+U Publishing Co., Ltd.
Publisher / Chief Editor: Nobuyuki Yoshida
Guest Editor: Channa Daswatte
Senior Editor: Kei Yokoyama Sylvia Chen
Editorial Staff: Grace Hong Ryosaku Onodera Ayako Sato
Oversea Assistant: HOU Lei

封面图：架构——伊玛杜瓦的别墅，建筑设计：帕林达·堪纳卡拉
封底图：2B1（集装箱住宅）建筑设计：阿米拉·德·梅尔

本书第272页至279页内容由安藤忠雄建筑研究所提供，在此表示特别感谢。

本系列丛书著作权归属文筑国际，未经允许不得转载。本书授权中国建筑工业出版社出版、发行。

Front cover: Interior view of The Frame – Holiday Home in Imaduwa.
Designed by Palinda Kannangara. Photo by Mahesh Mendis.
Back cover: Living space of 2B1 (Container House). Designed by Amila
de Mel. Photo courtesy of the architect.

The contents from pages 272 to 279 of this book were provided by
Tadao Ando Architect & Associates. We would like to express our
special thanks here.

The copyright of this series is owned by CA-GROUP. No reproduction
without permission. This book is authorized to be published and
distributed by China Architecture & Building Press.

Preface:
Architecture as Life or Architecture as a Discipline?

ZHAO Yang

In 2015, my friend C. Anjalendran invited me to Sri Lanka as a guest speaker at the Geoffrey Bawa Memorial Lecture. After the seminar, he gave me a tour of a series of houses and spaces he considered important and worth experiencing, several of which were his own work. What delighted me particularly was the meeting space at the end of the third-floor walkway in an urban private house located in the Columbus' Villa Rotunda. In this peaceful, elegant, and joyous space filled with the breath of life, I vaguely re-experienced the freedom that Bawa created through his effortless spatial arrangements. The architect did not seem to have been burdened by the architectural language; he simply arranged the facilities and scenes of life. We were invited by two rows of sofas facing each other, where we sat and had a conversation. A gentle light and images of green that entered from the windows surrounded us. There were almost no abstract forms to be seen here. Everything visible was what as it is in life: the white lacquered wooden windows were fitted with brass bolts, and the concrete frame beams were painted dark brown to match the teak rafters and roof boards. The furniture had no distinct shapes; they were just ordinary furniture that make one feel comfortable. If Alain de Botton asked me to choose a scene for his *The Architecture of Happiness*, I would think of this little living space.

Anjalendran also invited me to stay at his house and studio in Colombo's suburbs and told me about people living in Bawa's time as well as his own story. He mentioned that he had recently designed a small house next to the studio for his butler so that his family could reunite in Columbo. When I visited Anjalendran again in 2019 with my family, this cozy little house had filled with sheen of life. Then I thought of my own life in which I had to run around, and I actually felt envious of Anjalendran's lucky butler.

During this visit, I was fortunate to meet two other architects, Channa Daswatte and Amila de Mel, and the photographer Dominic Sansoni. They were figures who were immersed in the life and culture of a time represented by Bawa. There is a calmness within them that goes beyond the convention. Amila de Mel's Container House *(p. 222)* included in this issue is a manifestation of such an attitude. The way they live and their work seem to extract vitality directly from the life world, and they can hardly be defined by any specific architectural discourse.

In this special volume of Sri Lankan contemporary architecture commemorating the 100th birth anniversary of Bawa, we see the exploration undertaken by the younger generations of Sri Lankan architects and scholars in pursuit of professionalism and architecture. These works deeply penetrate the very reality of the island country, explore the position of Sri Lankan architecture in contemporary global architecture, and enthusiastically participate in the universalized architectural discussion. I greatly admire their sincerity. However, I am also regretful that the harmony between architecture and life seen in Bawa's work has become rarefied. Bawa's works do not have a strong sense of "manifesto" in contrast to those of most emerging young architects and even lack a sense of architectural discipline in a certain degree. The change is probably directly associated with the contemporary and dramatic transformation brought upon the society and lifestyle of Sri Lanka. I could feel the honesty of the new generation of architects toward their own contextuality and the contemporary world. Bawa probably did not have the pressures of "professionalism" and "professionality," which the architects today have to withstand. Perhaps the influence of the strong tradition represented by Bawa is precisely what the new generation of architects want to escape. In any case, it has been 100 years since Bawa's birth, and Sri Lanka's architectural culture remains full of vigour. The new generation of architects and the contemporary architecture of this nation are revealing their charisma in ways different from the great master Bawa.

序言：
是作为生活的建筑，还是作为"建筑学"的建筑？

赵扬

2015 年，我的朋友老安（C. Anjanlendran）邀请我作为当年杰弗里·巴瓦纪念讲座的演讲者访问斯里兰卡。讲座过后，他带我参观了一系列他觉得重要的房子或者值得体验的空间，其中也包括他自己的几个作品。在位于科伦坡圆厅花园的城市私宅里，那个位于三楼流线尽端的会客空间让我欣喜不已。在这个充盈着生活气息的，祥和、优雅而雀跃的室内，我又依稀感到巴瓦空间里那举重若轻的自由。建筑师在这里似乎没有建筑语言的负担，而仅仅是安排生活的设施和场景而已。两排对峙的沙发邀请我们坐下来，促膝交谈，并被柔和的光线和窗外的绿意所包围。这里几乎看不到抽象的形式，一切可见之物不过就是生活中如其所是的物品——白漆木窗、黄铜插销；混凝土框架梁刷成深棕色，为的是跟柚木椽子和望板颜色相和。家具也没有特别的形式感，只是一些让人倍感舒适而亲切的普通家具而已。如果阿兰·德波顿（Alain de Botton）请我为他那本《幸福的建筑》选一个场景，我大概会想到这个小小的会客厅。

老安还邀请我去他在科伦坡近郊的私宅和工作室小住，跟我讲巴瓦那个年代的人物和他自己的故事。他还提到他刚在工作室旁边为照顾他生活的男仆设计了一个小住宅，让他们一家人可以在科伦坡团聚。2019 年当我带着家人再次造访老安，这个温馨的小住宅已经被生活浸润出光泽。联想到自己生活的奔波，我竟也羡慕起老安这位幸运的男仆。

不仅仅是老安，那次有幸见到的另外两位建筑师，钱纳·达斯瓦特（Channa Daswatte），阿米拉·德·梅尔（Amila de Mel），还有摄影师多米尼克·桑索尼（Dominic Sansoni），这些曾浸淫于巴瓦所代表的那个年代和生活文化的人物，都有一种不落窠臼的从容透脱。这本书里收录的阿米拉·德·梅尔（Amila de Mel）的作品"集装箱住宅（Container House）"（见本书第 222 页）就是这种态度的体现。他们的生活和创造似乎都直接从大地吸取生气，却很难被建筑学的话题所限定。

这本为纪念巴瓦诞辰 100 周年的斯里兰卡当代建筑专辑，让我们看到更年轻一辈的锡兰建筑师在职业化道路和建筑学诸多方向的探索，这些作品深入岛国自身现实，探索斯里兰卡的建筑在当代世界中的位置，并积极地参与到全球化的建筑学讨论，其诚恳精进令人钦佩。但是另一方面，我也惋惜于巴瓦作品里那种融化到生活中的和一品质却越来越稀薄。和书里涌现的大多数新生代相比，巴瓦的建筑其实并没有那么强烈的"宣言"感，甚至没有太多"建筑学"的味道。这大概跟斯里兰卡当代社会和生活方式的剧烈变迁直接相关，我也能感觉到新生代建筑师面对自身处境和当代世界的诚实，建筑师以"职业性"和"专业性"立身，巴瓦当年大概没有类似的压力，或许巴瓦所代表的强大传统正是新生代建筑师迫切想要挣脱的影响。无论如何，100 年来，斯里兰卡的建筑文化仍然充满朝气，新生代建筑师和这个国家的当代建筑正在以不同于大宗师巴瓦的方式展现着他们的魅力。

ZHAO Yang
Principal, Zhaoyang Architects

赵扬
赵扬建筑工作室主持建筑师

Contemporary Sri Lanka on Geoffrey Bawa's 100th

Preface:
Architecture as Life or Architecture as a Discipline?   6
ZHAO Yang

Essay:
Before and After the End of the Beginning   14
Sean Anderson

32
Chapter 1: Materiality and Process
Thisara Thanapathy
Santani Wellness Resort and Spa   34

Palinda Kannangara
The Frame – Holiday Home in Imaduwa   50

Robust Architecture Workshop
Ambepussa Community Library   62

Philip Weeraratne
Kotte Residence   76

Robust Architecture Workshop
House 412   82

Palinda Kannangara
Studio Dwelling at Rajagiriya   92

104
Chapter 2: Conservation and Critical Reuse
Pulasthi Wijekoon, Guruge Ruwani, Thusara Waidyasekara
Office Building for Colombo Municipal Council   106

Thisara Thanapathy
TRACE Expert City   120

Dilum Adikari
Arcade Independence Square   130
Dutch Hospital Shopping Precinct   142

Channa Daswatte
Galle Fort Hotel   152

Amila de Mel
No.5 @ Lunuganga (Osmund and Ena de Silva House)   162

174
Chapter 3: An Environmental Ethic
Sunela Jayewardene
Jetwing Vil Uyana   176

L. A. R Kumarathunge, Tom Armstrong
The Mudhouse   188

Hirante Welandawe
Villa Santé   196
Jaffna House   204

Channa Daswatte
Daswatte House   212

Amila de Mel
2B1 (Container House)   222

teaM Architrave
Office Building for Central Finance Co. Plc   234

Essay:
When is a Contemporary Sri Lankan Architecture?   242
Sean Andersona and Channa Daswatte

Architects Profile   246

From the notebook of Shuji Kondo:
Memories of Geoffrey Bawa   250
Shuji Kondo

Bawa 100:
"The Gift" : Artist-panel Discussion   264
Suhanya Raffel, Dominic Sansoni, Lee Mingwei, Shayari de Silva, Chandragupta Thenuwara, Dayanita Singh, Sean Anderson, Christopher Silva

Spotlight:
Liangzhu Center of Arts   272
Tadao Ando Architect & Associates

斯里兰卡：
当代与巴瓦 100 周年

序言：
是作为生活的建筑，还是作为"建筑学"的建筑？　7
赵扬

论文：
黎明前的黄昏　14
肖恩·安德森

32
第一章：物质性与过程

蒂萨拉·泰纳帕里
圣塔尼温泉疗养度假村　34

帕林达·堪纳卡拉
架构——伊玛杜瓦的别墅　50

罗伯斯特建筑工作室
安贝普瑟社区图书馆　62

菲利普·韦拉拉特尼
科特公寓　76

罗伯斯特建筑工作室
住宅 412　82

帕林达·堪纳卡拉
拉贾吉里亚的建筑师寓所　92

104
第二章：保护和批判性再利用

普拉斯丁·维耶空，古鲁格·鲁瓦尼，苏萨拉·威迪亚塞卡拉
科伦坡市议会办公楼　106

蒂萨拉·泰纳帕里
TRACE 专家城　120

迪卢姆·阿迪卡里
独立广场购物中心　130
荷兰医院旧址购物美食广场　142

钱纳·达斯瓦特
加勒古堡酒店　152

阿米拉·德·梅尔
卢努甘卡第五号住宅（奥斯蒙德与埃娜·德·席尔瓦住宅）　162

174
第三章：环境伦理

苏妮拉·贾瓦德
杰特维茵·维尔·乌亚那度假酒店　176

L.A.R 库马拉通加，汤姆·阿姆斯特朗
泥屋　188

希兰特·韦兰达维
桑特别墅　196
贾夫纳之家　204

钱纳·达斯瓦特
达斯瓦特之家　212

阿米拉·德·梅尔
2B1（集装箱住宅）　222

阿奇特瑞弗建筑师事务所
中央金融有限公司办公楼　234

论文：
斯里兰卡建筑何时走向当代？　242
肖恩·安德森，钱纳·达斯瓦特

建筑师简介　246

近藤秀次 记：追忆杰弗里·巴瓦　250
近藤秀次

巴瓦 100 周年：
"礼物"：艺术家圆桌论坛　264
苏安雅·华菲，多米尼克·桑索尼，李明维，
沙耶里·德·席尔瓦，昌德拉古萨·特努瓦拉，达亚妮塔·辛格，
肖恩·安德森，克里斯托弗·席尔瓦

特别收录：
良渚文化艺术中心　272
安藤忠雄建筑研究所

Editor's Words

编者的话

Following *a+u* 11:06 issue on Geoffrey Bawa that featured the architect's masterpieces with his lasting influence over generations, this book examines the works of these emerging, new Sri Lankan architects. Beginning with an essay written by Sean Anderson, Associate Curator in the Department of Architecture and Design at MoMA, it describes the scene of how early practices by pioneering architects cope with the political economy of colonialism at the time to bring about modern regional architecture of the tropics. It is not until the year 1969, Sri Lanka started their own local architecture schools which taught new generations of architects who continue to challenge the limits of tradition and shape their country's architecture identity. Together with Channa Daswatte, architect at the MICD Associates and Trustee of the Geoffrey Bawa Trust and the Lunuganga Trust, 19 projects are selected and structured into 3 topics that reflect the architecture of contemporary Sri Lanka relevant to its economic, social, cultural situation today. (a+u)

a+u曾在2011年6月的斯里兰卡专辑里收录了建筑家杰弗里·巴瓦的一些代表作品，并讲述了巴瓦带给后辈的影响。本书中，a+u将继续聚焦斯里兰卡，介绍新一代建筑师的作品。在正文开篇，由美国现代艺术博物馆建筑与设计系副馆长肖恩·安德森先生撰写的文章中，我们将会了解斯里兰卡的先驱建筑家的早期作品是如何应对当时的殖民主义政治经济的，以及他们如何为世界带来了被称为"热带现代主义"学派的近代地域建筑。1969年，斯里兰卡首次设立了建筑院校。之后，他们不断挑战传统界限，逐渐诞生了新一代的建筑师，他们共同塑造了斯里兰卡建筑的身份认同。本书的客座主编是钱纳·达斯瓦特(MICD 协会、杰弗里·巴瓦财团理事长、卢努甘卡财团理事)。本书收录的19个作品将分三个部分进行介绍，以此追踪受经济、社会和文化现状影响的当代斯里兰卡建筑。

(a+u)

Projects List
项目一览

01　**Santani Wellness Resort and Spa**. Werapitiya, Kandy
　　圣塔尼温泉疗养度假村，康提，维尔佩第亚
02　**The Frame – Holiday Home** in Imaduwa, Baliyagoda Mulana, Wahala Kananke, Imaduwa
　　架构——伊玛杜瓦的别墅，伊玛杜瓦，瓦哈拉·卡南克，巴里亚果达·穆拉纳
03　**Ambepussa Community Library**, Sri Lanka Army, Ambepussa
　　安贝普瑟社区图书馆，安贝普瑟，斯里兰卡军队基地
04　**Kotte Residence**, Sri Jayawardenapura Kotte
　　科特公寓，科特，斯里·查亚沃登内布拉
05　**House 412**, Pannipitiya, Colombo
　　住宅412，科伦坡，潘尼皮蒂亚
06　**Studio Dwelling at Rajagiriya**, Buthgamuwa Road, Rajagiriya
　　拉贾吉里亚的建筑师寓所，拉贾吉里亚，布加穆瓦路
07　**Office Building for Colombo Municipal Council**, Dr C.W.W Kannangara Mawatha, Colombo
　　科伦坡市议会办公楼，科伦坡，C.W.W 堪纳卡拉·玛瓦莎博士路
08　**TRACE Expert City**, Maradana, Colombo
　　TRACE 专家城，科伦坡，马拉达纳
09　**Arcade Independence Square**, Independence Ave., Colombo
　　独立广场购物中心，科伦坡，独立大道
10　**Dutch Hospital Shopping Precinct**, Hospital St, Colombo
　　荷兰医院旧址购物美食广场，科伦坡，医院路
11　**Galle Fort Hotel**, Galle Fort, Galle
　　加勒古堡酒店，加勒，加勒古堡
12　**No.5 @ Lunuganga (Osmund and Ena de Silva House)**, Lunuganga, Dedduwa, Bentota
　　卢努甘卡第五号住宅（奥斯蒙德与艾娜·德·席尔瓦住宅），本托塔，德都瓦，卢努甘卡
13　**Jetwing Vil Uyana**, Sigiriya
　　杰特维茵·维尔·乌亚那度假酒店，锡吉里耶
14　**The Mudhouse**, Paramakanda, Anamaduwa, Puttalam
　　泥屋，普塔勒姆，阿讷默杜沃，帕拉马坎达
15　**Villa Santé**, Kappalady, Talawila, Kalpitiya
　　桑特别墅，卡尔皮蒂耶，塔拉维拉，卡帕拉迪
16　**Jaffna House**, Thirunavely, Jaffna
　　贾夫纳之家，贾夫纳，蒂鲁内尔维利
17　**Daswatte House**, Madiwela, Sri Jayawardenepura Kotte
　　达斯瓦特之家，科特，斯里·查亚沃登内布拉，曼迪维拉
18　**2B1 (Container House)**, Mirihana, Nugegoda
　　2B1（集装箱住宅），奴各果达，密里哈纳
19　**Office Building For Central Finance Co. Plc**, 268 Vauxhall Street, Colombo
　　中央金融有限公司办公楼，科伦坡，沃克斯豪尔路 268 号

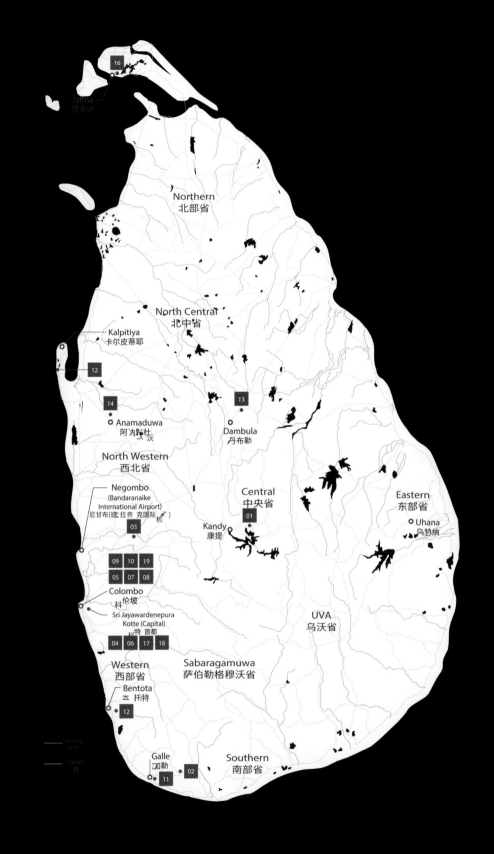

Essay:
Before and After the End of the Beginning
Sean Anderson

论文：
黎明前的黄昏
肖恩·安德森

"To be contemporary is to understand all of one's past, all of one's inheritance, and, while joining this to the present, to look towards the future…"

Kamlesh Dutt Tripathi
As quoted in *The Twice Born: Life and Death on the Ganges*, Aatish Taseer, 4th estate (Harper Collins), India, 2018

"In my personal search I have looked into the past for the help that previous answers can give, and at the pointers of previous mistakes. By the past I mean all the past, from Anuradhapura to the latest finished building in Colombo, from Polonnaruwa to the present moment – the whole range of effort, the peaks of beauty and simplicity and the deep valleys of pretension"

Geoffrey Bawa
Times of Ceylon Annual, Colombo, 1968

For its thousands of years of maritime history, Sri Lanka has been at the nexus of social, political and economic systems that today threaten the inheritance of an architecture and urbanism built for and in a fragile yet fecund island landscape. Over the past few years, the expansion of the island's urban contexts has mirrored that of the degradation of its seaboard and inland byways not only because of environmental instability but also because of the misguided promise of tourism as a sustainable economic model. The country's strategic importance in the South Asia-Indian Ocean region speaks to an outgrowth of investment in reshaping Sri Lanka's infrastructure. What began as a mode of connecting people in and across the island has also become a channel to distribute land ownership, to shift material resources and to link local corporations to networks of global capital. Consequently, the rapidly fluctuating skyline of the nation's commercial capital, Colombo, is testament to a frenzy of construction that commenced with the cessation of a thirty-year civil war in 2009.

How the built environment of Sri Lanka today can also be apprehended through its villages, their discrete organization and individual structures, is increasingly difficult. The integrity of small-scale building as a means of sociability does not match that of vast development occurring throughout the island. Architectural and urban projects that once were to illustrate an island's rebirth after the war are fundamentally changing the ways in which transportation, housing and sensitive environmental conditions are being rendered. In 2012, for instance, the opening of a new system of wide-laned expressways alongside other infrastructural projects including a new airport and shipping terminal was leveraged with access to international loans, the nonpayment of which led to the transfer of some of these facilities to the lenders. Similarly, the acquisition of a massive landfill extension to Colombo's traditional seaport was paid for by giving long leases on a significant portion of the reclaimed land itself. Contemporary architecture and urbanism, accordingly, has become one measure by which to understand not only the demands placed on civil society at a time

"所谓现代，就是要明白过去的所有，承接遗产的全部，并在将它们与现在结合起来的同时，展望未来……"

卡姆利什·杜特·特里帕蒂
引用于阿蒂西·塔西尔，《两次出生：恒河上的生与死》
第四地产（哈珀·柯林斯），印度，2018年

"当我思索难题时，我从过去的答案和错误中学习。所谓过去，即过去的全部建筑实践和尝试，从阿努拉德普勒到科伦坡最新建成的建筑，从波隆纳鲁沃到此时此刻，其中有美与天真之巅峰，也有骄傲自负之深渊。"

杰弗里·巴瓦
《锡兰时报年刊》，科伦坡，1968年

作为一个岛国，斯里兰卡拥有数千年的航海历史，一直处在社会、政治和经济体制动荡的中心，这也是今日斯里兰卡脆弱但成果斐然的岛屿景观建筑和城市建造传承仍有可能遭受破坏的原因。在过去几年中，都市扩张和海陆环境恶化同时发生，这不仅缘于环境的不稳定性，还因为错信了旅游业是一种可持续的经济发展模式。对于重建斯里兰卡基础设施的投资源源不断，这是基于该国在南亚、印度洋地区的战略重要性。这些建设原是帮助岛上民众相互往来的手段，但在此基础上渐渐发展成了分配土地所有权、转移物质资源和将本地公司与全球资本网络联系起来的渠道。商业首都科伦坡的迅速发展，反映了2009年斯里兰卡长达三十年的内战结束以来，猛烈的建设狂潮。

如今，想通过村庄的内部组织以及个人建造活动来理解斯里兰卡的人居环境已越来越难。一座小型建筑虽然可以满足所有社会需求，但不能代表整个岛屿的巨大发展。这些曾经象征战后重生面貌的建筑和城市项目正在从根本上改变着交通设施、住房建设和环境保护的整体图景。比如在2012年，斯里兰卡通过海外贷款，用举债经营的方式启用了新机场、运输航站楼等基础设施以及配套的数条高速公路，但由于无力偿还国际债务，最终不得不将其中一些设施转让给贷方。同样地，在扩张科伦坡海港土地的大规模填埋项目中，资金大部分来源于长期出租填海土地的租金。因此，现代建筑与城市规划被综合成了一种方法，用来在世界不统一的时代中了解公民的社会需求，同时理解斯里兰卡规划和设计为何以及如何适应改变。

自从加勒大道和新开发的滨海大道上林立了密集的、由开发商主导的酒店和休闲商业设施，曾经在科伦坡市中心就能够观赏到的寻常海景如今已经不复存在。而建于填海土地之上的港口城也面临同样的问题。这些商业项目几乎都由国外建筑师设计，导致了城市空间密度与占地面积的不平均，这在市中心及其周围区域最为明显（第17页，图1）。在科伦坡这座历史悠久的中心城市中，混居着不同阶层的人群，而这些新项目致使其中大量的贫困人口流离失所。大规模的房地产投机买卖不仅在科伦坡兴起，还波及了远方的加勒，人们摒弃空间的共享，转向了经济和阶级差异明确区分的模式。新自由主义的玻璃幕墙包裹着配有空调的内部空间，空间内排列着一连串冷冰冰的监视器。如字面意思那样，曾经在大海另一边可以看到的海岸线被分割，消失在西式、中东式、东南亚式混合的海市蜃楼中。然而，许多人正是被这种虚假的形象所吸引（第17页，图2）。

这种翻天覆地的变化正在世界各地发生。在今天全球化高度关联却有着自毁倾向的社会背景下，我们不得不

of disunity in the world but also the how and why sensitive planning and design in Sri Lanka should accompany change.

Sea frontage, once a common, taken-for-granted occurrence in downtown Colombo, will no longer be possible given the spate of developer-led hotel, leisure and commercial complexes that are now lined up along the Galle Road, the newly-opened Marine Drive and soon in the Port City development built on reclaimed land. Almost always designed by architects from outside the country, such facilities now contribute to unequal density and inequitable occupations that are most noticeable in and around the city center (*p. 17, Image 1*). Some of the projects have been responsible for the displacement of large economically disadvantaged urban populations who lived within the palimpsest of economic strata found in the historical center. The rise of mass real estate speculation in Colombo and further afield in Galle suggests a turning away from shared spaces toward a model of highly defined economic and class differences. Framing a series of unresponsive sentinels in this neoliberal cascade of glass curtain-wall facades with their air-conditioned interior realms, the horizon, once seen across a wide ocean – both literally and figuratively – has been disassembled and rebuilt in the guise of a hybrid Western-Middle Eastern and Southeast Asian mirage. And it is this invented image to which many are drawn (*p. 17, Image 2*).

Such unmitigated development, happening throughout the world, demands the question: What is it to be contemporary today in the context of a globalized, hyperconnected yet self-destructive planetary society? For Sri Lanka, the response is a complex one that requires an observance of problematic colonial histories while also being attuned to the nature of the local architectural profession, its pedagogies, its ways of working with spatial adjacencies and materials to attend to the present and the past simultaneously. Like the definition proposed by the Brahmin priest about being contemporary, successful contemporary architecture in Sri Lanka may be one that embodies that which has gone before, while ascribing such conflicted legacies to an outlook predicated on confidence in the future.

It follows then to ask, what and how is the past understood in relation to Sri Lankan contemporary architecture? How does one understand that the adaptation of ideas over time can become the basis of a responsive yet unique architecture? In the absence of an established methodology of construction or the blurring of traditions with using materials in a specific way, pedagogy is one mode by which the narratives of construction, of buildings as both signals and images, are transmitted from one generation to another.

The first architecture school in the country was founded as an extension of Bauhaus principles that remained an aesthetic currency within architectural and artistic pedagogies well into the 1990's. It is only in the last decade of the twentieth century that one finds the work of architects born and working in Sri Lanka from the mid-20th century onward introduced as part of the education of architects. Within the two schools of architecture, the Colombo University School (founded in 1969) and the newly created Sri Lanka Institute of Architects School of Architecture (now the City School of Architecture – Colombo, founded in 1986), what was conceived as contemporary was a shifting target for both instructors and students alike.

Colonial spatial histories across South Asia were observed as contradictory to the ideals of Socialist thinking prevalent among the political élite from the 1960s onward. For Sri Lanka, architectural design pedagogies came to fruition between the 1960-1980s, condensing in some part, the nomenclature and structuring of Western modernism. Unlike Jawarhalal Nehru's bringing Le Corbusier and numerous other Western-trained architects and designers to India, in addition to the convergence of aid-based architecture projects in West Pakistan as instruments for asserting newness, the official policies in Sri Lanka eschewed this attention toward the aesthetics of radical materialities such as reinforced concrete and steel. Rather, in 1948, the government of the former Ceylon deployed traditional Sri Lankan architectural motifs to camouflage classically-designed buildings as seen in Shirley de Alwis's

This page: Image 1, High-rise buildings along the streets of Colombo. Photo by Kasun Jayamanne. Image 2, View of Colombo Skyline. Photo by Banuka Vithanage. Image 3, Peradeniya University Senate Building designed by Shirley de Alwis. Photo courtesy of the University of Peradeniya. Image 4, Karunaratne House designed by Minette de Silva.

本页，图1：科伦坡街道上的高层建筑。图2：科伦坡天际线。图3：雪莉·德·阿尔维斯设计的佩拉德尼亚大学参议院大楼。图4：明奈特·德·席尔瓦设计的卡鲁纳拉特纳住宅。

University of Ceylon at Peradeniya, one of the first large building projects constructed in that year (*Opposite, Image 3*). This merging of recognizable pre-colonial historical forms with new institutional buildings became a hallmark of a Sri Lankan architecture that looked in two directions.

The island's intersection with an externally-defined modern was initially established through local practitioners whose education and training abroad led to instances of innovative domestic architecture. Built from 1949–51 in Kandy, the Karunaratne House (*Opposite, Image 4*) by Minnette de Silva was an assured departure from the colonial bungalow typology. A multi-level plan affords ample spaces for entertaining while also ensuring a distinct separation of private spaces and service areas. From its façade to its use of glass block and undulating walls, the concrete and stone house with its hilltop view of the city's sacred center and the Temple of the Tooth, is anachronistic for the period yet seductive in the tension between inside and outside, the visible and the phenomenal.

In an article written in 1953, "A House in Kandy," in an issue of the newly founded Bombay-based Marg magazine, Minette de Silva argues for a climatically and culturally relevant modern architecture for South Asia. De Silva, the only Asian woman CIAM participant, who later became an editor at the magazine, also published her Karunaratne House in the same issue. For both the house, and by extension, Sri Lankan architecture, she discusses how cultural traits and climate must begin to modify the core ideas of modernity while adapting to a particular culture and place. While adhering to the tenets of an interconnected modernism, including "indigenous methods of construction," De Silva made concerted efforts through her work, particularly in conceiving of new prototypes in housing such as the Senanayake Flats in Colombo (1954–57, *this page, Image 5*). For her project of a hotel in Sigiriya and the later Coomaraswamy house in Colombo (1974, *this page, Image 6*) she examines the adaptation of local vernacular forms while reconceptualizing them as a basis for contemporary lifestyles. Prominent Sri Lankan art historian Ananda Coomaraswamy had advocated such affinities years prior. For De Silva, and throughout her storied career, the marriage of new materials with prevailing regional building traditions allowed for a syncretic application of local skills and crafts to establish an architecture that was closely related in time and place.

Histories of shared experience, both personal as well as communal, became a source of inspiration when making new buildings. While Sri Lankan government policies in the 1960's and early 1970's encouraged self-reliance with a goal toward controlling economic outflows, architects were forced to contend with balancing aesthetic experimentation and political realities. To understand the work of the early Sri Lankan modernists, two patterns emerged: on the one hand, those that proposed an architecture that deployed the standard materials of mechanization, evocations of the International Style. On the other, how to work within existing social, political and fiscal frameworks to counter overt foreign intervention. To build, for both, was a means

提出一个问题:"现代化"到底是什么?斯里兰卡的答案是十分复杂的,在同时处理新旧时代的情况下,不仅需要考虑饱经风霜的殖民历史,还要理解当地建筑业与教育的本质,邻近空间与当地材料的处理方法。就像文章起始那位婆罗门僧所谈及现代的定义那样,斯里兰卡的现代建筑若想成功,应该承接历史,将这充满矛盾的遗产放置于对未来发展充满信心的前景之上。

那么,接下来的问题是:在斯里兰卡现代建筑中,"过去"是什么?它又是如何被理解的?如何使人们领悟,随时间推移而改变的思想与观念,可以成为兼备适应性和独特性建筑的基石?在缺乏约定俗成的建造方法,或混018传统材料使用方法的情况下,教育是可以将建筑与建造的故事作为标志和图像,传给下一代的。

斯里兰卡的第一所建筑院校继承了包豪斯的教育方针。在20世纪90年代之后,包豪斯的美学主张仍是建筑和艺术教育的主流。直到20世纪的最后十年,20世纪中叶以来斯里兰卡建筑师在本土的作品才开始出现在建筑教育体系中。在科伦坡大学建筑学院(成立于1969年)和稍晚一点的斯里兰卡建筑师学会建筑学院(现为城市建筑学院,1986年成立于科伦坡市)里,教师和学生共同构想、探讨"现代建筑",但仍未能对其定义达成定论。

整个南亚殖民地的空间历史,与自20世纪60年代以来政治精英阶层中盛行的社会主义理想之间存在矛盾。在斯里兰卡,建筑设计教育在20世纪60年代至20世纪80年代间取得的成果一定程度上凝聚了西方现代主义的话语和体系。不同于贾瓦哈拉尔·尼赫鲁将勒·柯布西耶和许多其他受西方体系训练的建筑师和设计师带到印度,也不同于带到巴基斯坦西部聚集的新社会援建项目,斯里兰卡官方避开了以钢筋混凝土和钢材等为代表材料的激进美学。实际上,在1948年,旧锡兰政府就运用传统斯里兰卡建筑图案伪装经典来设计建筑,如雪莉·德·阿尔维斯设计的位于佩拉德尼亚的锡兰大学(现为佩拉德尼亚大学)。这也是那年建造的第一批大型建筑项目之一(第17页,图3)。这种将可识别的前殖民历史形态与新公共建筑融合的方法成为斯里兰卡建筑两面兼顾的标志。

斯里兰卡最先产生的外界所谓的"现代建筑",是学成归来的斯里兰卡建筑师设计的新式本土住宅。明奈特·德·席尔瓦于1949-1951年间设计的位于康提的卡鲁纳拉特纳住宅(第17页,图4),完全从殖民时期的平房类型中跳脱解放。多层平面为娱乐功能提供了充足的空间,同时确保私人空间和服务区域的泾渭分明。玻璃砖立面、波浪形墙体、混凝土与石料的构建,以及可以俯瞰城市神圣中心和牙之圣殿的山顶景观,在当时或许有些过时,但在内与外、视觉与感官之间产生了无比的张力与魅力。

1953年,明奈特·德·席尔瓦为当时刚成立的孟买杂志《玛格》撰写了《康提的一座住宅》一文,文中他提倡建造符合南亚气候和文化的现代建筑。德·席尔瓦是近代建筑国际会议(CIAM)中唯一一位亚洲女性建筑师。她设计的卡鲁纳拉特纳之家也发表于上述同一期刊中,后来她成为《玛格》的编辑。她认为斯里兰卡的住宅乃至整个建筑界不必盲从于从国际引入的"现代主义",而应该根据文化和气候特质来理解现代主义的核心思想,并且随着特定区域和文化适应改变。德·席尔瓦将包括"本土建造方法"的现代主义宗旨贯彻于实践中,并积极与他人合作。这样的设计态度在构思新住宅原型,如科伦坡的森纳那亚克公寓(1954–1957,第18页,图5)时,尤为明显。德·席尔瓦在锡吉里耶的酒店项目和之后在科伦坡的库玛拉斯瓦米之家项目(1974年,第18页,图6)中研究了本地乡土建筑形态现代化的可能性,将重构的概念作为当代生活方式的基石。斯里兰卡著名艺术史学家阿南达·库玛拉斯瓦米早

Opposite: Image 5, Senanayake Flats Colombo designed by Minette de Silva in 1954. Image 6, Coomaraswamy House designed by Minette de Silva in 1974. Images 4–6 courtesy of L&WoAWA. Photo courtesy of Monocle, Issue 65, Part 02 Colombo.

对页,图5:明奈特·德·席尔瓦于1954年设计的科伦坡森纳那亚克公寓。图6:明奈特·德·席尔瓦于1974年设计的库玛拉斯瓦米之家。

of suggesting potential while also attempting to resolve the problems of an emergent nation state. Construction materials, however, such as steel and glass, both of which were imported into the country (along with rice, ironically), suddenly had severe restrictions placed on them in terms of price as well as usage. Epitomized among the several buildings made for the 1965 Industrial Exhibition held in central Colombo, (*This page, Image 7*) the Planetarium designed by the engineer A.N.S. Kulasinghe with the architects Kandavel and Panini Tennakoon working in the Public Works Department is one example of daring structural definition merging with a then-rarely used material technique, prestressed concrete. Such ideals of autonomy led the government and cadres of financiers and manufacturers to ensure a resurgence of cultural specificity erupt in many aspects of society.

One finds among the work of Valentine Gunasekera and Panini Tennakoon as well as the early work of Geoffrey Bawa, translations of modernism inspiring new forms. In his House for the Illangakoon (1969, This page, Image 8) and the Tangalle Bay Hotel (1972, Opposite, Image 9) Gunasekera assays concrete and glass as a bulwark against the Sri Lankan tropical climate. By comparison, Tennakoon's National Archives building (1972, Opposite, Image 10) reminiscent of interwar Italian architecture, achieves monumentality through its expressive arched colonnade and finned concrete façade. At the same time, the landscape for these projects was redefined not as a component of the architecture itself but as a stage upon which the buildings assimilated the

This page: Image 7, Planetarium designed by the engineer A.N.S. Kulasinghe with architects, Kandavel and Panini Tennakoon in 1965. Photo courtesy of the author. Image 8, Illangakoon House designed by Valentine Gunasekera in 1969. Opposite: Image 9, Tangalle Bay designed by Valentine Gunesekera in 1972. Image 10. National Archives building designed by Panini Tennakoon in 1972. Photo courtesy of Monocle, Issue 65, Part 02 Colombo. Images 8, 9, 22, 23 courtesy Geoffrey Bawa archive.

本页，图 7：工程师阿鲁玛杜拉·南达塞纳·席尔瓦·库拉辛格与建筑师坎达维尔和帕尼尼·特纳库恩于 1965 年共同设计的天文馆。图 8：瓦伦丁·古那瑟卡拉于 1969 年设计的伊兰加贡之家。对页，图 9：瓦伦丁·古那瑟卡拉于 1972 年设计的唐加勒湾酒店。图 10：帕尼尼·特纳库恩于 1972 年设计的国家档案馆。

在几年前就倡导这两者之间的密切联系。在她整个传奇职业生涯中，德·席尔瓦将当地盛行的建筑传统与新材料融合，同时运用本土技艺，建造了与时间和地点紧密相连的建筑。

共享的个人体验和社会共有的历史经验成为新建筑的灵感来源。20世纪60年代至70年代初期，为应对斯里兰卡政府拉动内需和控制资本外流的政策，建筑师被迫在美学实验和政治现实之间寻求平衡。早期现代主义建筑师的态度大概可归为两种：第一种基于机械化生产的标准材料进行设计，建筑作品响应国际风格；另一种则坚持在既有社会、政治和财政的体系之内开展工作，以对抗国际主义的明显介入。对于这两者而言，唯有建造，既是表明潜力的方式，又能尝试解决新兴国家的问题。在斯里兰卡，钢和玻璃等建筑材料一直依赖进口（具有讽刺意味的是，还有大米）。它们的进口突然受到严厉管制，导致价格上涨，用途受限。1965年科伦坡市中心工业展览会（对页，图7）中具有代表性的建筑是天文馆，由工程师阿鲁玛杜拉·南达塞纳·席尔瓦·库拉辛格与公共工程部建筑师坎迪维尔和帕尼尼·特纳库恩共同设计。该建筑采用了大胆的结构与当时罕见的预应力混凝土材料技术。政府领导、金融家和制造商力图通过这种自治的典范，促使斯里兰卡文化特质在社会各方面复苏。

瓦伦丁·古那瑟卡拉、帕尼尼·特纳库恩和杰弗里·巴瓦的早期作品都受到现代主义的启发，在转译基础上发展出新的形式。古那瑟卡拉在他设计的伊兰加贡之家（1969年，对页，图8）和唐加勒湾酒店（1972年，本页，图9）中使用混凝土和玻璃，试验将它们用作抵御斯里兰卡热带气候的屏障的可能性。相比之下，特纳库恩设计的国家档案馆建筑（1972年，本页，图10）则会使人联想到内战期间的意大利建筑。富有表现力的拱形柱廊和肋状混凝土立面构成了一座纪念性建筑。同时，这些项目的景观被重新定义，它们不再是建筑的一部分，而是新的舞台。当内外兼修，并与斯里兰卡的自然景观相融合时，现代主义精神所孕育的南亚区域主义便萌芽了。

最早在斯里兰卡推广适应南亚气候的现代建筑观念的是简·德鲁和麦克斯韦·弗莱。他们关于跨国热带建筑的构想是在伦敦建筑协会建筑学院工作室中建立的。弗莱团队在整个西非和其他新兴后殖民城市的项目，对全球范围内的现代施工概念（施工和外观方面）产生了影响。同时，这些项目也成为将气候融入建筑构造的参照。弗莱和德鲁的开创性著作《干旱和潮湿地区的热带建筑》（1964年）将杰弗里·巴瓦设计的圣托马斯学院（1957-1964）视为通过动态空间设计应对气候环境的一个案例（第22页，图11）。上述项目和其他建筑，例如帕尼尼·特纳库恩设计的科伦坡新议会大厦（1974年，第22页，图12），几乎都讨论度较低。但从与过去诀别的角度看，它们与尼赫鲁委托柯布西耶在昌迪加尔和艾哈迈达巴德建造的作品，以及路易斯·康在达卡、理查德·纽特拉在卡拉奇、康斯坦丁诺斯·多克迪什在伊斯兰堡建造的作品何其类似。

当巴瓦尝试弗莱和德鲁倡导的现代主义本土化时，他在某种程度上也受到了新同僚乌尔力克·普莱斯纳的影响。这位丹麦建筑师及工程师曾为明奈特·德·席尔瓦工作过，并受其信念鼓舞，致力于为斯里兰卡建造适宜当地的建筑。巴瓦和普莱斯纳很快便开始了探索——如何使用本土建筑材料来应对当时的政治环境。巴瓦早期的住宅项目，例如加勒的A.S.H.德·席尔瓦住宅（1959-1960年，对页，图13）和作为范式的埃娜·奥斯蒙德与德·席尔瓦住宅

This page: Image 11, St. Thomas College designed by Geoffrey Bawa in 1957. Image 12, New parliament designed by Panini Tennakoon in 1974. Photo courtesy of the heirs of Panini Tennakoon. Image 13, A.S.H. de Silva House designed by Geoffrey Bawa in 1960. Image 14, Ena de Silva House designed by Geoffrey Bawa in 1960. Opposite: Image 15, Steel Corporation Offices designed by Geoffrey Bawa in 1966. Image 16, Bentota Beach Hotel designed by Geoffrey Bawa in 1968.

本页，图11：杰佛里·巴瓦于1957年设计的圣托马斯学院。图12：帕尼尼·特纳库恩于1974年设计的科伦坡新议会大厦。图13：杰弗里·巴瓦于1960年设计的A.S.H.德·席尔瓦住宅。图14：杰弗里·巴瓦于1960年设计的埃娜·德·席尔瓦住宅。对页，图15：杰弗里·巴瓦于1966年设计的钢铁公司办公楼和宿舍。图16：杰弗里·巴瓦于1968年设计的本托塔旅游度假村。

(1960—1962年，对页，图14），尊重了居住者所处的城市环境，有着内向性的空间布局且造价相对便宜，而且在对待建造和形态的态度上以体验为核心。学者香提·贾瓦德认为，巴瓦试图通过批判殖民建筑遗产来超越过去（香提·贾瓦德，《杰弗里·曼宁·巴瓦——非殖民化巴瓦》，国家信托基金会，2017年）。

在接下来的岁月中，杰弗里·巴瓦成了斯里兰卡最负盛名和最高产的建筑师之一。巴瓦的代表作，例如位于奥鲁瓦拉科伦坡郊外的钢铁公司办公室和宿舍（1966—1969年，本页，图15）、本托特旅游度假村（1967—1969年，本页，图16）和当时岛上最高的建筑之一，12层的国家信托银行（1976—1978年，第24页，图17）全部采用高度创新的混凝土框架，从而实现自然通风、开阔空间和充足光线。这些项目使西方学者将巴瓦视为在南亚实现新兴"批判性地域主义"的先驱。但是这种由西方学术界定义并推动的"主义"是值得怀疑的。讽刺的是，曾经也正是这群人抹杀了德·席尔瓦、古纳瑟克拉和巴瓦等建筑师的作品，他们试图印证在美国和西欧之外出现现代主义并非不可能，却也是不正常的。

20世纪80年代之前是公认的西方建筑风格霸权时期，强调混凝土和其他工业建筑生产形式的概念。此后，本土化和自给自足的设计概念更加被重视。从德·席尔瓦、古那瑟卡拉和特纳库恩的各种作品中都可以观察到这种注重材料的建造实践，但这样的态度在斯里兰卡似乎很快便消失了。1985年伦敦英国皇家建筑师协会（RIBA）举办的巴瓦作品展和1986年出版的《巴瓦——斯里兰卡的建筑师》，确立了巴瓦的现代主义批评家和敏锐地方观察者的国际地位。但从巴瓦的角度来说，他只是在特定的政治、经济、文化和气候条件下，设计适用且实用的建筑（与钱纳·达斯瓦特的谈话）。

就地域性而言，材料主义及本土概念两种论述都对斯里兰卡的建筑实践产生了深远影响。巴瓦和古那瑟卡拉于1957年接管了爱德华兹、瑞德与贝格建筑师事务所。该事务所于科伦坡成立，是一所大型建筑公司。整个20世纪60年代，许多实习生来到该公司，学习广泛采用当地材料进行建造并尝试内外空间整体设计等概念，逐渐成长为下一代建筑师。在当时的建筑界，密切关注乡土建筑技法，同时考虑自然通风和采光的设计已逐渐被接受并开始流行。

20世纪70年代之后出现了一系列建筑空间形态排列的实验性作品。阿努拉·拉特纳维布山从1963年至1979年任职于巴瓦的事务所，他在莫拉图瓦设计的自宅（1982年）环绕倒映池而建（第25页，图18），使用了明显的现代主义手法，成了室内外整体设计的典范。瓦桑莎·雅各布森作为管理者催生了巴瓦的数个大型项目。她在几个住宅项目中展现了自己对于城市生活的看法，尝试将住宅从城市环境中剥离。她的作品关注空间与形态中的光影关系。雅各布森位于博雷拉的自宅（1979年）故意模糊了内外空间（第25页，图19）。另外，特纳·维克拉玛辛格在为国家工程公司工作时，把过去的经验融入到峰会公寓（对页，图20）的设计中。该公寓是一组砖砌建筑，为1976年在科伦坡举行的非同盟诸国首脑会议的各国记者提供住处。建筑师如何将景观与邻近的殖民政府住宅和谐融合，引发了殖民历史和现今广泛国际合作间的对话。对于上述每位建筑师而言，物质文化是一种通过经验来构成、定义和铭刻意义的载体。

20世纪80年代初期，斯里兰卡局势紧张，内战一触即发。为了方案能够平稳实施，大量建筑师尝试从历史构建

new. The fusion of Sri Lanka's natural landscape and a propensity to look within and outside at the same time, catalyzed the notion of a South Asian regionalism borne from a modernist ethos.

Arguments for a climatically responsive modern architecture in South Asia was promoted initially in Sri Lanka through the works of Jane Drew and Maxwell Fry. Their inferences concerning a trans-national tropical architecture were established in a studio at the Architectural Association School of Architecture in London. The team's work throughout West Africa and other emergent post-colonial centers was also influential in the conception of modern construction that was once international in scope (of building and appearance) but could also act as an instrument through and by which climate became part of architectural composition. Geoffrey Bawa's St Thomas College (1957–64) appears in Fry and Drew's seminal book, Tropical Architecture in the Dry and Humid Zones (1964), as one example of dynamically designing through climate (p. 22, Image 11). These and other buildings such as Panini Tennakoon's designs for the new Houses of Parliament in Colombo (1974, p. 22, Image 12) are rarely discussed examples of building in Sri Lanka that signaled a break with the past as much as Nehru desired in the work of Le Corbusier in Chandigarh and Ahmedabad; or Louis I. Kahn in Dhaka, and Richard Neutra in Karachi and Constantinos Doxiadis in Islamabad.

Contemporaneous with his attempts at localizing a modernism promoted by Fry and Drew, Geoffrey Bawa was also possibly influenced in part by the arrival in his office of Ulrik Plesner. The Danish architect and engineer, who had for a short time worked for Minette de Silva, had been encouraged by her ideas to conceive of approaches to making a relevant architecture for the country. Both Bawa and Plesner soon began to explore the use of local building materials as a way to respond to the political climate of the time. Evident in the Dr. A.S.H. de Silva House in Galle (1959-60, p. 22, Image 13) and the paradigmatic House for Ena and Osmund de Silva (1960–62, p. 22, Image 14) Bawa's early residential projects were respectful of their occupants in urban settings, inward-looking and relatively inexpensive, but also highly engaged examples toward an experientially centered attitude to building and form. The scholar Shanti Jayewardene contends that this approach was Bawa's attempt to make an intellectual break with the past through a critique of colonial building legacies. (Geoffrey Manning Bawa – De-Colonizing Bawa, Shanti Jayewardene, National Trust, 2017).

In the years that follow, Geoffrey Bawa became one of the most sought after and prolific architects in Sri Lanka. Bawa's key buildings such as the Offices and Housing for the Ceylon Steel Corporation on the edges of Colombo at Oruwala (1966–69, p. 23, Image 15) ,the Bentota Beach Hotel (1967–69, p. 23, Image 16) and one of the tallest buildings on the island at the time, the twelve-story State Mortgage Bank building (1976–78, this page, Image 17) all arrayed highly innovative structural concrete to allow for natural ventilation, ample spaces and access to light. Ironically, such projects propelled Bawa as having

Opposite: Image 17, State Mortgage Bank Building designed by Geoffrey Bawa in 1978. This page: Image 18, House in Moratuwa designed by Anura Ratnavibhushana in 1982. Photo from the collection of Anura Ratnavibhushana. Image 19, House in Borella designed by Vasantha Jacobson in 1985. Photo courtesy of the author.

对页，图 17：杰弗里·巴瓦于 1978 年设计的国家信托银行。本页，图 18：阿努拉·拉特纳维布山于 1982 年设计的莫拉图瓦住宅。图 19：瓦桑莎·雅各布森于 1985 年设计的博雷拉住宅。

enabled a newfound "critical regionalism" in South Asia. Such was the questionable efficacy of "–ism's" defined by and propelled through western scholarship that once effaced the work of architects such as De Silva, Gunasekara and Bawa, while also ensuring that questions of modernism outside the bounds of the United States and Western Europe were unusual, if not impossible.

After several years of what may be regarded as a western hegemony of architectural styles, by the 1980s, the adoption of localized and self-sufficient approaches to design outweighed conceptions of concrete and other forms of industrial building production. Such evocations of materialist approaches to building in Sri Lanka, found among the varied works of De Silva, Gunasekara, and Tennakoon, appear to have soon faded. In 1985, an exhibition of Bawa's work at the Royal Institute of British Architects in London and the 1986 publication of Bawa – Architect in Sri Lanka secured his position internationally as both a critic of modernism as well as a keen observer of place. Bawa, on the other hand, simply took a position that he was building what was sensible and useful to the political, economic, cultural and climatic conditions in which he was working. (Private conservation with Channa Daswatte)

Such contrasting discourses concerning the intersection of locality with an overt materialism and concepts of indigeneity had a profound influence on the practice of architecture in Sri Lanka. Edwards, Reid and Begg, one of the larger offices established in Colombo was taken over by Bawa and Gunasekara in 1957. Due to succeeding generations of architects interning in the office throughout the 1960s, key concepts proffered by its associates included extensive building with local materials as well as attempts at the seamless integration of inside and outside. Designing with natural ventilation and light while paying close attention to autochthonous building practices became an accepted vocabulary that began to prevail in the architecture of the period.

Beginning in the 1970s, a distinct series of projects illustrated experiments in the choreography of spatial form. Having worked in Bawa's office from 1963 until 1979, Anura Ratnavibhushana's own

House at Moratuwa (1982) exemplifies a prismatic integration of interior and exterior while using an overtly modernist vocabulary set around a reflecting pond (p. 25, Image 18). Vasantha Jacobsen, herself a catalytic figure for managing a number of Bawa's larger projects, had her own views of urban living which manifested in several houses that sought to distinguish the house within an urban setting. Her work centered on bringing light and shade into buildings as both spatial and formal encounters. For her House in Borella (1979), the distinction between interior and exterior is made deliberately indistinct (p. 23, Image 19). By extension, while working for the State Engineering Corporation, Turner Wickramasinghe integrated multiple influences in his Summit Flats, a series of brick-finished apartments to house journalists of the 1976 Non-aligned conference held in Colombo (this page, Image 20). How the architect integrated a resonant landscape with adjacent colonial government residences induces a dialogue concerning memory and a cooperative present. For each of these

和当地景观中寻求突破。例如，1985年，伊斯梅尔·拉希姆和菲罗兹·乔克西测量并分析了公元前4世纪古遗迹里提加拉修道院的平面剖面，根据研究成果设计了哈巴拉娜旅馆酒店(对页，图21)。公共与私密，可见与不可见交织，他们于苍松森林间寻求修道院花园般与自然共存的平衡。

巴瓦于1976-1978年设计的，位于科伦坡中部贝拉湖的西马马拉卡寺(第28页，图22)受到喀拉拉邦寺庙建筑的启发，使用跨文化的设计方法来实现混合建筑特色。其他具有代表性的例子还有使用了相似类型学的科特巴特拉穆拉的林业局办公楼(第28页，图23)和其他建筑。那些为树立斯里兰卡建筑身份认同的建筑师们，从承接历史声音的阶段迈向了积极与外界交流的舞台。

博学多才的艺术家拉吉·塞纳纳雅克最初在爱德华兹、里德和贝格建筑师事务所工作，并于1970年隐居到丹布拉附近干旱地区的森林迪亚巴布拉，在那里搭建了露天凉亭。迪亚巴布拉这个地名源于当地的泉水，而塞纳纳雅克在此地逐渐建造了一系列可居住的建筑。这里成了塞纳纳雅克在艺术和建筑探索中，社会和空间伦理的具象化，并展示了如何用质朴的手法服务于现代生活(第29页，图24)。同样地，巴瓦的亲密伙伴C.安贾伦德兰在他1989年设计的SOS青年村中，将佛教寺院的庭院形态纳入了丰富多彩的开放式住宅中，进一步促进当地空间范式的应用(第29页，图25)。毫无疑问，这样的空间结构只能由熟知并且可以引导、记录整个岛屿本土建筑和纪念性建筑发展的建筑师塑造。

整个1970年代，在德国、英国等大规模基础建设项目的影响下，耶德沃德政府领导的"马哈维利发展计划"制定了新的城镇规划，并承诺为100万农民提供住房。乌尔里克·普莱斯纳负责该计划的一部分。在他1967年离开巴瓦事务所后，被召回斯里兰卡担任工程顾问，并成为马哈维利建筑部的部长。他的工作还包括特尔登尼亚等城镇的那些由于水坝建设被淹没的重建项目(第29页，图26)。1977年，当斯里兰卡从国家主导的政体开放为资本主义自由市场之后，大量新材料和新商品涌入；同时，人们更容易到国外旅行并接触到国际出版物，从而产生了新的思潮。多年来，这些对于绝大部分国民来说是一种几乎享受不到的奢侈。可以说，这样的开放带来了建筑思想的百花齐放与蓬勃发展。然而，由政府发起的大型项目追随了那些世界上最早民主议会国家的方向和形态。大规模的公共项目，例如1979-1982年建设的国民议会大楼(第29页，图27)和南部一所占地广阔的新大学鲁胡努大学(1980-1988年，第31页，图28)都委托给了巴瓦。这种对现代建筑地域化的拥护成为了斯里兰卡设计和建设的模式，并且作为一种形式语言广为人知。相比之下，千禧年之际国家建筑管理局设计的许多官方建筑都体现了所谓的斯里兰卡"民族主义"独特风格(第31页，图29)。

如今的互联网世界，"热带现代主义"被与高级的生活方式混为一谈，并被理解为异域化，以及斯里兰卡建筑表现力的下降。那些受杰弗里·巴瓦影响和曾在巴瓦事务所工作过的建筑师们，在质疑过去与环境价值的同时，各自发展着自己的建筑语言。尽管受到许多外部影响，或者也正是因为这些影响，斯里兰卡的现代建筑形象并不清晰。准确地说，现如今各个规模的建筑都有巴瓦和早期现代主义建筑师的影响。针对稳定气候的设计和建造，也不仅仅出现在特定的观光设施中。此外，在当今斯里兰卡的施工条件下，斯里兰卡建筑师在进口材料及外来形式，与本地材料及传统形式的不协调与矛盾之间进行选择和抗争。钢结构与赤陶瓦屋顶、地砖巧妙地结合在一起，铝框玻璃窗与素混凝土地板组合，再加上废弃的建筑构件，它们既回应了环境，又对所处社会加以注解。这些对建筑构造的考量，正是斯里兰卡及其周边诸国的建筑师们努力的结果，他们努力解决大量流动人口的需求，回应他们转瞬即逝的幻想与欲望。

在远离高密度城市和村庄的边缘，矗立着一座座用于国际贸易的混凝土塔楼；偏远的海滨和丛林幽静的住所中，有着属于斯里兰卡建筑师们的职责和理想。我们在慰藉中勇敢，在光影中表达；最终，在过去和现在的叙事中，建筑的约束与力量拒绝着任何简单的定义。

Opposite: Image 20, Summit Flats designed by Turner Wickramasinghe in 1975. Photo by Varuna de Silva. Image 21, Habarana lodge designed by Pheroze Choksy and Ismeth Raheem in 1985. Photo courtesy of Cinnamon Hotels.

对页，图20：特纳·维克拉玛辛格于1975年设计的峰会公寓。图21：菲罗兹·乔克西和伊斯梅尔·拉希姆于1985年设计的哈巴拉娜旅馆酒店。

architects, material culture was a way to compose, define and inscribe meaning through experience.

By the early 1980s, with tensions foreshadowing the beginnings of civil war, a profusion of architects working across the island sought historical structures as well as landscapes as a way to steadfastly situate their proposals. In 1985, for example, Ismeth Raheem and Pheroze Choksy analyzed the measured plans and sections of monastic ruins found at the ancient 4th century hermitage of Ritigala and arranged the accommodation of their Habarana Lodge Hotel (p. 26, Image 21) based on their findings. The structuring of public and private spaces, the seen and unseen, amidst a forest of mature trees sought to achieve a concomitant balance as that found within a monastic garden.

Directly inspired by Keralan temple architecture, the Seema Malaka (This page, Image 22) on Beira lake in central Colombo designed by Bawa from 1976–78, began to suggest trans-cultural approaches toward hybrid building characteristics. Such is evident with the use of typological forms at the Offices of the Forest Department in Battaramulla in Kotte (This page, Image 23) and other buildings. Forging an identity for a Sri Lankan architecture pointed to the purveyance of historical messages merging with an outward, almost gregarious view to living.

Artist and polymath, Laki Senanayake, who first worked in the offices of Edwards, Reid and Begg, retreated to his own open pavilion set in the dry zone forest near Dambulla in 1970. Named for the water spring he found there, Diyababula was organized over time as a series of habitable structures. The site became a model of social and spatial ethics that Senanayake deployed in his art and architecture to show how contemporary life could be lived within modest means (p. 29, Image 24). Similarly, C. Anjalendran, who was a close associate of Bawa, in his SOS Youth Village of 1989, adopted the geometries of Buddhist courtyard monasteries into colorful open plan residences that furthered the use of local spatial paradigms (p. 29, Image 25). Here, the structuring of spaces was no doubt informed by the architect's extensive knowledge, teaching and documentation of vernacular and monumental architecture found throughout the island.

Throughout the 1970s, the planning of new townships affected by the construction of massive infrastructural projects managed by countries such as Germany and the U.K., were developed by the Mahaweli Development Scheme under the Jayewardene government who also promised homes for one million farmers. Managed in part by Ulrik Plesner, having been recalled to Sri Lanka as an engineering consultant after he left Bawa's practice in 1967, as head of the Mahaweli architectural unit, towns such as Teldeniya were designed to replace those that had been submerged with the building of dams (p. 29, Image 26). The opening of the Sri Lankan economy from its hitherto direct model to free market Capitalism in 1977 saw the influx, not only of new materials and goods for consumption, but also a free flow of ideas influenced by easier access to foreign travel and publications. For many years, these luxuries had been impossible for the majority of

Opposite: Image 22, Seema Malaka designed by Geoffrey Bawa in 1978. Image 23, Offices of Forest Department of Sri Lanka Photo by Kasun Jayamanne. This page: Image 24, Diyabubula designed by Laki Senanayake in 1970. Photo by Dominic Sansoni. Image 25, SOS youth village designed by C. Anjalendran in 1989. Photo from the collection of C. Anjalendran. Image 26, Mahaweli New town designed by Ulrik Plesner in 1979. Image 27, Parliament designed by Geoffrey Bawa in 1979.

对页，图 22：杰弗里·巴瓦于 1978 年设计的西马马拉卡寺。图 23：卡桑·贾亚曼于斯里兰卡拍摄的林业局办公楼。本页，图 24：拉吉·塞纳纳雅克于 1970 年设计的迪亚巴布拉。图 25：C·安贾伦德兰于 1989 年设计的 SOS 青年村。图 26：乌尔里克·普莱斯纳于 1979 年设计的马哈维利新镇。图 27：杰弗里·巴瓦于 1979 年设计的斯里兰卡新议会大厦。

the population and such openings arguably resulted in a plethora of architectural ideas flourishing. Other major government-initiated projects, however, established the direction and image that one of the world's first democratic parliamentary governments had chosen. Massive public projects such as the National Parliament buildings from 1979–82 (*This page, Image 27*) and the University of Ruhunu campus (1980–88, *Opposite, Image 28*) a vast new university in the southern part of the island, were given to Bawa. Such allegiances to a locally sourced modern architecture established a mode of design and building that is recognizable for its formal attributes. In contrast, by the turn of the millennium, a number of the official buildings designed by the State Building Authorities embody what some have described as a distinct Sri Lankan "nationalist" style (*Opposite, Image 29*).

The Internet view of the world, having conflated the idea of a "tropical modernism" with that of lifestyle branding, further problematizes what may be understood as both an exoticization as well as a diminishing of architectural expression in Sri Lanka. Those architects that had been influenced by or had worked in the studios of Geoffrey Bawa, developed their own architectural language while interrogating valences of the past and its environments. In spite of these external influences or perhaps because of them, the appearance of a Sri Lankan contemporary architecture today is not necessarily self-evident. Rather, the influence of Bawa and the early modernist architects appears at multiple scales. Sensitive approaches to designing and building in response to unvaried climatic flows need not only be expressed in an architecture for a particular kind of tourism. Further, while considering domestic construction today, one may contend with a seeming contradiction between the use of imported materials and forms alongside those that appear within the country. Steel structures are deftly combined with terracotta tile roofs and floor tiles; glass windows framed in aluminum are tooled with plain concrete floors; disused old building elements are reused both as an environmentally sensitive approach as well as a comment on entropy. These architectonic entanglements speak to contemporary architects in Sri Lanka and those looking on from other countries throughout the region wrestling with the demands of an engaged mobile populace and fleeting images of desire.

What one finds, beyond the densification of its cities and village edges, past the concrete towers that intensify global commerce, and among isolated beachside and jungle retreats, is an architecture that asserts the divergent role and ambition of Sri Lankan architects. We are introduced to boldness in solace, expression in light and shadow, and ultimately, in whose narratives crossing past with present, the restraint and power of an architecture that resists definition.

References:
1. Maria Abi-Habib. How China got Sri Lanka to cough up a port, New York Times, NewYork: June 25, 2018
2. Geoffrey Bawa. A way of Building, Times of Ceylon Annual, Colombo, 1966
3. Channa Daswatte. Bawa on Bawa in Tan Kok Meng, Asian architects 2, Singapore: Select Books, 2001
4. Minnette de Silva. A House in Kandy, MARG Vol.6 No.3, Bombay: June 1953
5. Minnette de Silva. Experiments in Modern Regional architecture in Ceylon, Journal of the Ceylon Institute of Architects, Colombo: 1965/66
6. Minnette de Silva. The life and work of an Asian woman architect, Vol.1, Colombo: Smart Productions, 1998
7. Jane Drew and Maxwell Fry. Tropical architecture in the dry and Humid zones, London: Batsford, 1964
8. Shanti Jayewardene. Geoffrey Manning Bawa – De-Colonizing Bawa, Colombo: The National Trust of Sri Lanka, 2017
9. Oxford Business Group. Sri Lanka's Construction sector sees increased investment and Development, The Report – Sri Lanka, www.oxfordbusinessgroup.com: 2016
10. Anoma Pieris. Imagining Modernity, Pannipitiya: Stamford Lake Pvt Limited, 2007
11. David Robson. Bawa - The complete works, London: Thames and Hudson, 2002
12. David Robson. Beyond Bawa, London: Thames and Hudson, 2007
13. Aatish Taseer. The Twice Born: Life and Death on the Ganges, New Delhi: 4th estate, Harper Collins, 2018

Opposite: Image 28, University of Ruhuna designed by Geoffrey Bawa in 1985. Image 29, Hambantota New Secretariat designed by UDA. Photo courtesy of the Hambantota Secretariat.

对页，图 28：杰弗里·巴瓦于 1985 年设计的鲁胡努大学。图 29：UDA 设计的汉班托塔新秘书处。

Sean Anderson is Associate Curator in the Department of Architecture and Design at the Museum of Modern Art. He has practiced as an architect and taught in Afghanistan, Australia, India, Italy, Morocco, Sri Lanka and the U.A.E. His second book, In-Visible Colonies: Modern Architecture and its Representation in Colonial Eritrea (2015) was nominated for an AIFC Book Prize in Non-Fiction. At MoMA, he has organized the exhibitions "Insecurities: Tracing Displacement and Shelter" (2016–17), "Thinking Machines: Art and Design in the Computer Age", 1959–89 (2017–18) and manages the Young Architects Program (YAP) as well as the "Issues in Contemporary Architecture" series, with an exhibition concerning spatial justice and the American city planned for Autumn 2020.

肖恩·安德森是纽约现代艺术博物馆建筑与设计部门的助理策展人。他是执业建筑师，并在阿富汗、澳大利亚、印度、意大利、摩洛哥、斯里兰卡和阿联酋任教。他的第二本书《不可见的殖民地：现代建筑及其在厄立特里亚殖民地的表现》（2015年）获得了"AIFC非小说类图书奖"的提名。在纽约现代艺术博物馆，他策展了"不安全感：追踪流离失所和庇护所"（2016–2017年），"思维机器：计算机时代的艺术与设计1959–1989"（2017–2018年），并负责青年建筑师计划（YAP）以及"当代建筑问题"系列。并曾计划在2020年秋季举办有关空间正义和美国城市的展览。

参考文献：
1. 玛丽亚·阿比·哈比卜《中国与斯里兰卡的汉班托塔港交易》《纽约时报》，纽约，2018年6月25日
2. 杰弗里·巴瓦《建筑之道》《锡兰时报年刊》，科伦坡，1966年
3. 钱纳·达斯瓦特《巴瓦——陈国孟的巴瓦》《亚洲建筑师2》，新加坡：精选集，2001年
4. 明奈特·德·席尔瓦《康提的住宅》《玛格》第6卷第3期，孟买，1953年6月
5. 明奈特·德·席尔瓦《锡兰的现代地方性建筑实验》《锡兰建筑师学院期刊》，科伦坡，1965-1966年
6. 明奈特·德·席尔瓦《一位亚洲女性建筑师的作品和一生 第一卷》，科伦坡：智慧出品，1998年
7. 简·德鲁和麦克斯韦·弗莱《干燥与潮湿区域的热带建筑》，伦敦：巴斯佛德，1964年
8. 香提·贾瓦德《杰弗里·曼宁·巴瓦——非殖民化巴瓦》，科伦坡：国家信托基金会，2017年
9. 牛津商业集团《斯里兰卡的建筑迎来了更多的投资和发展》，报告——斯里兰卡，www.oxfordbusinessgroup.com，2016年
10. 阿诺玛·皮利斯《想象现代》，潘尼皮蒂亚：斯坦福湖私营有限公司，2007年
11. 大卫·罗布森《巴瓦作品全集》，伦敦：泰晤士和哈德森，2002年
12. 大卫·罗布森《超越巴瓦》，伦敦：泰晤士和哈德森，2007年
13. 阿蒂西·塔耶尔《两次生来：恒河上的生与死》，新德里：第四地产，哈珀·柯林斯，2018年

Chapter 1:
Materiality and Process

第一章：
物质性与过程

From its first modern incarnations on the island, material has not only been a component of building but also a means by which to identify changing attitudes toward climate, structure and form. The ascendance of concrete, for instance, was first seen in Colombo as early as 1938 when the Baurs Flats were built with engineering from the Swiss bridge designer Maillart. Andrew Boyd's work with concrete, especially his house in Kandy from 1940, is an example of residential design that confounded builders with its modernist vocabulary. During the immediate post-independence period, the use of imported materials such as concrete and glass signaled shifting aesthetic tastes inasmuch as access to new markets. Minnette de Silva's reading of art historian Ananda Coomaraswamy, however, brought into focus a hybridization of international forms with a strong responsibility toward using local crafts and materials. For Geoffrey Bawa, a modern house could be built using entirely local materials with the minimal use of concrete. Among these architects, working with and responding to locality, also manifested desire for various segments of society was as much a political act as it was financially sound.

Sri Lanka is at a crossroads today with easier access to materials not only from the region but also from throughout the world. The distribution of resources and wealth remains difficult however when considering contemporary architecture. For whom is the design and building with local materials and crafts of critical importance? Read broadly, materiality and materialism are not mutually exclusive. Techniques of making architecture in Sri Lanka, as in decades past, is as much about the sharing of skills as about furthering the meaning of materials, their economic value and ultimately, a logic borne from the crossing of image into form.

Sean Anderson

在斯里兰卡，自现代主义出现以来，材料便被认为不仅仅是建筑的组成部分，还是一种用来识别人们对气候、结构和形式的态度的方法。例如，斯里兰卡最早认识到混凝土材料的优越性，要追溯到位于科伦坡的鲍尔公寓楼（1938年），它由瑞士桥梁设计师罗伯特·马亚尔设计建造。而安德鲁·博伊德的混凝土作品，尤其是他1940年在康提设计的住宅，则由于其现代主义语言而使当地建筑商们感到疑惑。在接踵而至的后独立时代，随着新市场的开拓，进口材料（例如混凝土和玻璃）逐渐被使用，这也标志着岛国斯里兰卡的审美品位在发生转变。然而，明奈特·德·席尔瓦对艺术史学家阿南达·库马拉斯瓦米的解读，将国际形式与使用当地工艺材料的责任相结合，引起了人们的关注。对于杰弗里·巴瓦而言，一座现代住宅完全可以使用当地材料和少量混凝土来建造。这些回应地域性的建筑师表现出了对社会发展各个方面的愿景，这既是一种政治行为，也有其经济上的合理性。

如今，斯里兰卡正处于既可从所在区域，也能从世界各地轻松获取材料的十字路口。但从现代建筑的角度看，资源和财富的分配仍然十分困难。使用本地材料和手工艺品进行设计和建造，到底对谁而言是至关重要的？广义地讲，物质性与物质主义并不相互排斥。今天，斯里兰卡的建造技艺与过去几十年一样，都是关于技术的分享、材料深层意义、经济价值以及最终将图像转化为形式的逻辑。

肖恩·安德森

Thisara Thanapathy
Santani Wellness Resort and Spa
Werapitiya, Kandy 2014–2015

蒂萨拉·泰纳帕里
圣塔尼温泉疗养度假村
康提，维尔佩第亚 2014–2015

The design of this wellness center comprises two main elements, a spa complex and a residential complex. They are both accessed through a common entrance pavilion. The architecture is designed to relate to, and disappear into the landscape – embracing the spiritual tranquility of its surroundings. The carefully framed views, play of light, capture of cool breezes, and the use of rustic natural materials along with steel and glass, create a multi-sensory experience that leads to a natural sense of relaxation. The entrance pavilion, lying at the center of the valley, anchors the complex within the landscape by defining a clear vista across it and by framing scenic views on both sides. From the pavilion entrance, one of the two sides lead to the residential part of the complex dominated by the two-story Lounge-Restaurant – situated on the highest point of the land, embracing the picturesque mountain scenery. The 16 lightweight accommodations – resting on steel pillars – are arranged in a cascading manner down the sloping landscape, similar to traditional huts often seen on terraced paddy fields in the hills, all placed to capture the stunning views of the distant hills, terraced fields, and forests below.

In the opposite direction, the entrance leads to the Spa. The reception pavilion of the spa, inspired by Kandyan vernacular structures, is a lightweight timber structure resting on pillars, raised slightly above ground. From the reception, follows a seemingly underground series of rooms – with a play of levels – connected through naturally top lit tunnels and corridors made of stone; these treatment rooms have highly defined views that evoke a feeling of peaceful captivity.

Text by Thisara Thanapathy,
edited by Channa Daswatte

Credits and Data
Project title: Santani Wellness Resort And Spa
Client: Dumbara Hotels Pvt. Ltd.
Location: Werapitiya, Kandy, Sri Lanka
Design: 2014
Completion: 2015
Architect: Thisara Thanapathy Architects
Design Team: Thisara Thanapathy (principal), Rafidh Rifaadh, Kaushala Samarawickrema
Project Team: Wasantha Chandrathilaka (structural engineer); Sunanda Gnanasiri (quantity surveyor); Thilak Thembiliyagoda, Nimal Perera (MEP engineer); Duminda Builders Pvt Ltd (civil contractor)
Project area: 4,645 m^2
Project estimate: USD 1,930,000 (340 million Sri Lankan Rupees)

pp. 34–35: The transparency of the restaurant building merges into its surroundings, creating a subtle connection with nature. Photo courtesy of the architect. Opposite, above: The single villas rest lightly on the mountain capturing its serene surrounding views. Photos on pp. 36–49 by Thilina Wijesiri unless otherwise noted. Opposite, below: The entire resort, made up of lightweight structures, aligns itself atop the mountain ridge without interfering much of the natural habitat. Photo by Marielle Lindah.

第34-35页：有通透感的餐厅融入周围环境，与自然形成了微妙的联系。对页，上：坐落于山上的单栋别墅，可以将周围的静谧景色尽收眼底；对页，下：整个度假村的建筑都由轻型结构建成。它们沿山脊线而建并且不会干扰野生环境。

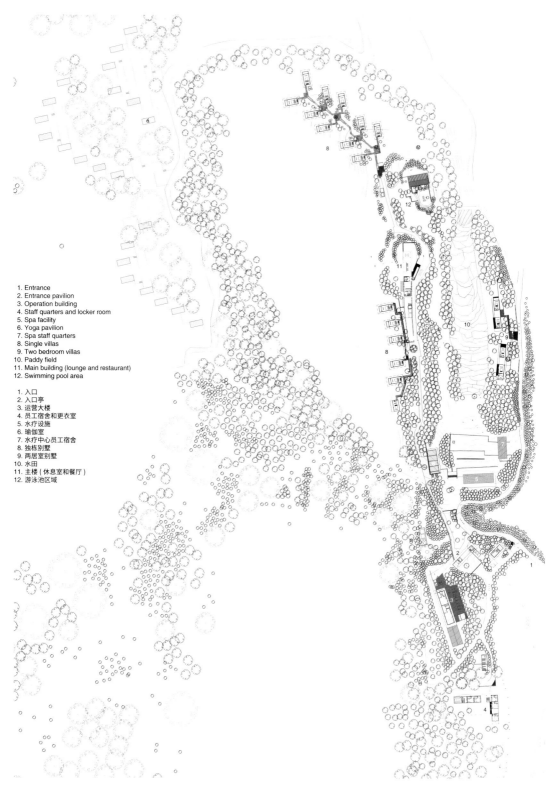

Site plan (scale: 1/3,000) /总平面图（比例：1/3,000）

1. Entrance
2. Entrance pavilion
3. Operation building
4. Staff quarters and locker room
5. Spa facility
6. Yoga pavilion
7. Spa staff quarters
8. Single villas
9. Two bedroom villas
10. Paddy field
11. Main building (lounge and restaurant)
12. Swimming pool area

1. 入口
2. 入口亭
3. 运营大楼
4. 员工宿舍和更衣室
5. 水疗设施
6. 瑜伽室
7. 水疗中心员工宿舍
8. 独栋别墅
9. 两居室别墅
10. 水田
11. 主楼（休息室和餐厅）
12. 游泳池区域

圣塔尼温泉疗养度假村由水疗中心和住宅群两部分组成。它们共用一个亭子作为入口。该建筑的设计旨在连接并融入景观，拥抱环境所赋予精神上的宁静。精心定格的视野、光的嬉戏、吹拂而过的凉风、钢与玻璃还有天然材料的使用，会营造出多种感官的绝佳体验，并带来自在的放松感。入口亭坐落于山谷中心，能够把远景清晰地收入视野，并在两侧构筑优美的景观，将整个建筑群置于景观之中。从入口出发，两侧都有道路。其中一侧通向住宅区，那里有此地海拔最高的双层酒廊餐厅，被如画的山景所环抱。16个轻型住宅顺着山坡向下依次排列。它们坐落在钢柱上，就像丘陵梯田上常见的传统小屋那般。每个住宅都可以将远处丘陵、梯田和下方森林的壮丽景色尽收眼底。入口的另一侧则通向水疗中心。该中心的接待亭受康提乡土建筑高床式结构的启发，使用了轻巧的木结构并被柱子抬起，略高于地面。经过接待亭会来到一系列高低不等、看似处于地下的理疗室。它们通过有自然采光的石制隧道和走廊彼此相连。这些房间拥有清晰的视野，能够给人带来平静的感受。

蒂萨拉·泰纳帕里 / 文
钱纳·达斯瓦特 / 编

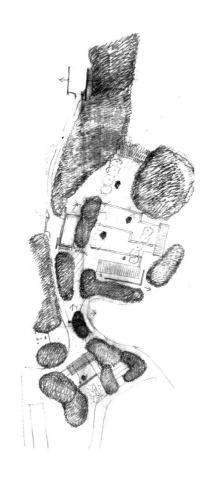

This page: Sketch of the spa with landscape. Image courtesy of the architect.

本页：水疗中心及其景观的手绘草图。

Opposite: View from the main building looking north towards the mountain ridges and greenery. This page: View of the restaurant above. The rhythm of the steel and timber members integrates the interior space with its external environment.

对页：从主楼向北眺望山脊和草木。本页：二层餐厅的样貌。钢材和木材构件将内部空间与外部环境融为一体。

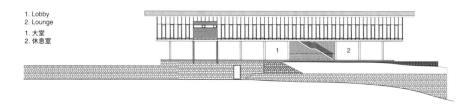

Long section of main building (scale: 1/350)／主建筑纵向剖面图（比例：1/350）

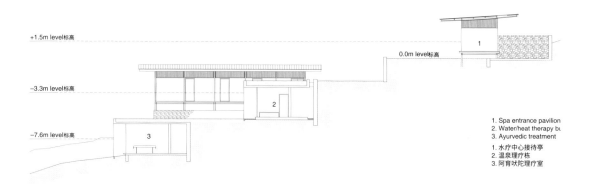

Long section of spa facility (scale: 1/350)／水疗设施纵向剖面图（比例：1/350）

pp. 42-43: The lobby of the main building opens up to the views of the natural landscape allowing for an immersive experience. p. 44: The entrance pavilion of the spa facility rests on top of a solid rubble base. p. 45, above: View of the relaxation lobby of the massage treatment facility. p. 45, below: The entrance lobby frames the view of the distant landscape as one approaches the spa facility. pp. 46–47: Dark enclosure of the hydrotherapy chamber capturing a glimpse of the scenery outside. Photo courtesy of the architect. This page: The textural quality of the materials is further highlighted with the play of light and shadows inside the spa facility.

第 42-43 页：主楼大堂向自然景观敞开，带来浸入式的体验。第 44 页：水疗中心的入口亭坐落于坚固的碎石基座之上。第 45 页，上：按摩设施所在的休闲大厅；第 45 页，下：当访客走向水疗中心时可以欣赏大堂入口处的美景。第 46-47 页：从幽暗的水疗室可以捕捉到一丝外面的景色。本页：水疗设施内部的光影活动进一步强调了材料的质感。

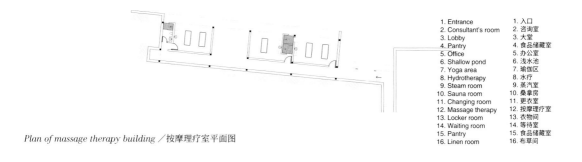

1. Entrance
2. Consultant's room
3. Lobby
4. Pantry
5. Office
6. Shallow pond
7. Yoga area
8. Hydrotherapy
9. Steam room
10. Sauna room
11. Changing room
12. Massage therapy
13. Locker room
14. Waiting room
15. Pantry
16. Linen room

1. 入口
2. 咨询室
3. 大堂
4. 食品储藏室
5. 办公室
6. 浅水池
7. 瑜伽区
8. 水疗
9. 蒸汽室
10. 桑拿房
11. 更衣室
12. 按摩理疗室
13. 衣物间
14. 等待室
15. 食品储藏室
16. 布草间

Plan of massage therapy building /按摩理疗室平面图

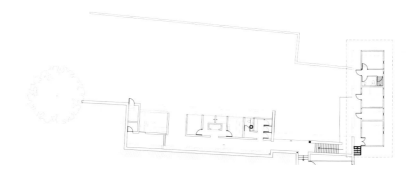

Plan of water and heat therapy building and workers' facility /温泉理疗室及员工设施平面图

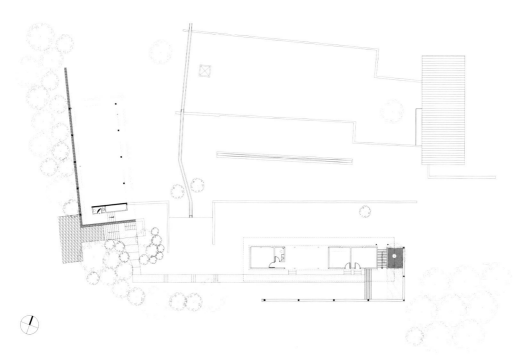

Plan of Entrance pavilion and Yoga pavilion (scale: 1/350) /入口及瑜伽室平面图（比例：1/350）

Palinda Kannangara
The Frame – Holiday Home in Imaduwa
Baliyagoda Mulana, Wahala Kananke, Imaduwa, 2018

帕林达·堪纳卡拉
架构——伊玛杜瓦的别墅
伊玛杜瓦，瓦哈拉·卡南克，巴里亚果达·穆拉纳，2018

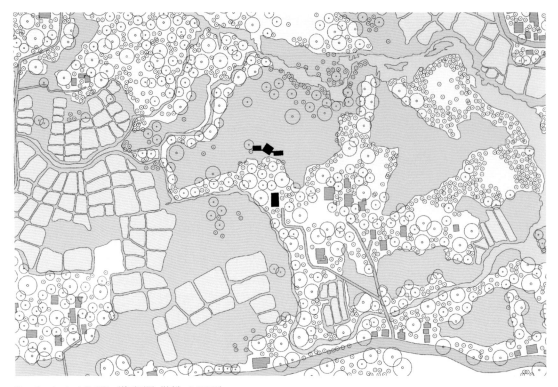

Site plan (scale: 1/5,000) /总平面图（比例：1/5,000）

pp. 50–51: Approach to the house. The entry staircase is made out of steel mesh and frame, while the boardwalk connecting between the rooms uses recycled scaffolding platform panels elevated on galvanised iron pipes. Opposite, above: The bedroom walls are lightweight cement board on timber frames, interspersed with glass louvers for light and ventilation. Opposite, below: Like an arthropod of steel scaffolding, the structure gently lifts the retreat off its terrain in the flood prone landscape. Photos on pp. 50–60 by Luka Alagiyawanna unless otherwise noted.

第 50-51 页：通向房屋的路径。入口楼梯由钢网和钢框架制成。连接房间之间的小道则使用回收而来的脚手架平台板，架在镀锌铁管之上。对页，上：卧室墙为铺在木框架上的轻质水泥板，其中点缀着玻璃百叶窗，以实现采光和自然通风；对页，下：就像由钢铁脚手架构成的节肢动物一样，该结构在水灾多发的地形中将假屋抬起。

This structure is a holiday home for a Sri Lankan ethno-musician and jazz drummer, and his family; it is built on ancestral lands that had been abandoned due to frequent floods. Located in southern Sri Lanka, about 6 km from Sinharaja Rainforest – a world heritage natural site – the structure negotiates a flood prone landscape, making minimal impact to the surrounding environment. The idea of the lightness and temporality of the landscape determined the decision to use steel scaffolding as an exoskeleton. Built on a small budget, the structure entirely made of component parts offers the flexibility to dismantle and reassemble this building on higher ground if desired. The structure comprises of 3 main platforms: a living, dining and pantry platform, and 2 other bedroom platforms – all connected together by light steel bridges. These platforms are angled to command views of the surrounding vegetation, creek and distant mountains; a simple belvedere looking out over a view.

Text by Palinda Kannangara,
edited by Channa Daswatte

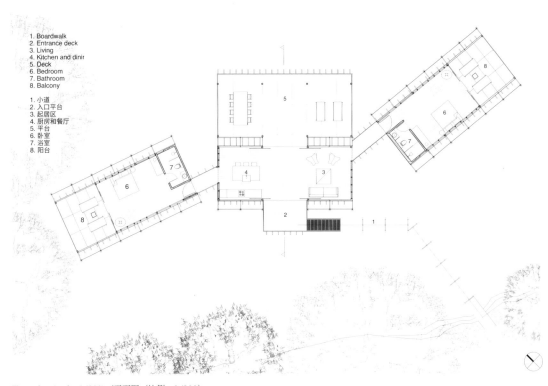

Floor plan (scale: 1/300) ／平面图（比例：1/300）

Credits and Data
Project title: The Frame – Holiday Home in Imaduwa
Client: Dr. Sumudi Suraweera (ethnomusician and Jazz drummer)
Location: Baliyagoda Mulana, Wahala Kananke, Imaduwa,
Sri Lanka
Design: 2018
Completion: Dec 2018 (construction duration: 4 months)
Architect: Palinda Kannangara Architects
Project Team: Ranjith Wijegunasekara (structural engineer)
Project area: 225 m²(built area), 32,375 m² (site area)
Project estimate: USD 40,000 (8 million Sri Lankan Rupees)

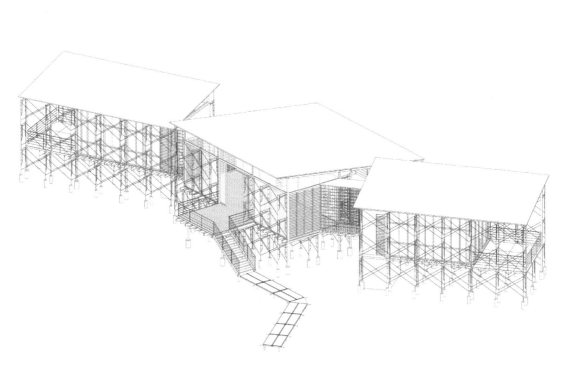

Axonometric drawing ／轴测图

这座建筑是一位斯里兰卡民族音乐人兼爵士鼓手和家人的度假屋，建于他们世代相传的土地上。这片土地曾因频发洪水而被遗弃。建筑位于斯里兰卡南部，距离世界自然遗产辛哈拉加雨林约6千米，设计旨在应对频发的洪水对景观的影响，同时最小化建筑对环境的影响。使用钢制脚手架作为外立面的灵感来自于当地景观的轻快和短暂。建筑建造的成本低廉，结构完全由预制构件组成，创造了拆卸和重组的灵活性，并可根据需要移建于更高的地面上。建筑由3个主要平台组成：起居、用餐及储藏平台和2个卧室平台。它们全部由轻钢桥架连接在一起。置身平台之上，可以欣赏周围的植被风景、小溪和遥远的山脉，俨然一座极简的眺望台。

帕林达·堪纳卡拉 / 文
钱纳·达斯瓦特 / 编

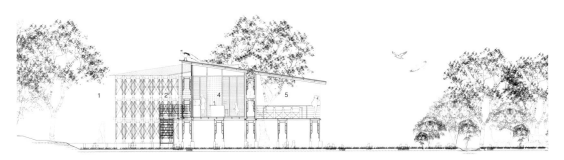

Section (scale: 1/350) ／剖面图（比例：1/350）

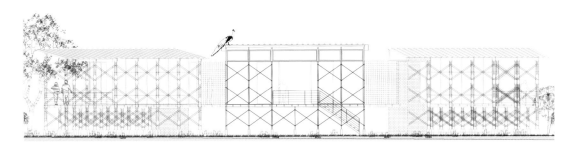

South elevation (scale: 1/350) ／南立面图（比例：1/350）

pp. 56-57: The outdoor living decks use steel checker plate flooring allowing its spaces to be permeable to views of the surrounding landscape. Photos on pp. 56-57, p. 61 by Mahesh Mendis. p. 58, above: View of the open verandah living and dining space. p. 58, below: Leftover wood from the decks are used in the construction of the kitchen that expands out into the living and dining area. p. 60: Construction mesh is used along the sides of the passage link between the living area and the bedrooms to provide a form of protection; and when necessary blinds can be rolled down to provide shading. p. 61: The floors of the bedrooms are made of repurposed timber supported on horizontal timber purlins, while the floorings of the verandahs use painted steel checker plates.

第56-57页：室外地板采用网纹钢板，使周围风景也渗透其中。第58页，上：开放阳台式的起居、就餐空间；第58页，下：甲板搭建所剩余的木材被用于建造延伸到起居和就餐空间的厨房。第60页：在起居区和卧室之间的通道两侧使用金属网，以提供保护。必要时可以将百叶窗卷下以提供遮蔽。第61页：卧室地板使用了回收而来的水平木材桁条，阳台地板则使用了涂漆的网纹钢板。

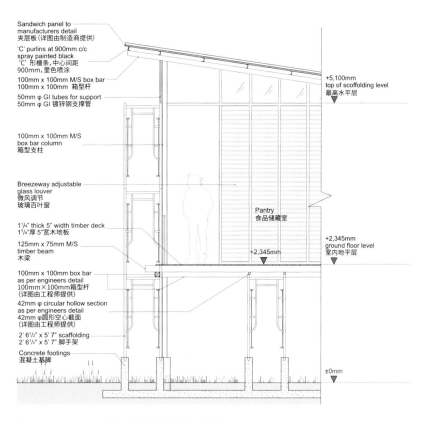

Detail section (scale: 1/75) /剖面详图（比例：1/75）

Robust Architecture Workshop
Ambepussa Community Library
Sri Lanka Army, Ambepussa 2013-2015

罗伯斯特建筑工作室
安贝普瑟社区图书馆
安贝普瑟，斯里兰卡军队基地 2013-2015

Site plan (scale: 1/12,000)／总平面图（比例：1/12,000）

pp. 62–63: The library meanders around existing landscape elements, creating indoor and outdoor spaces, framing views, and building formal references with the site. Opposite: The butterfly roof dances down the hill, establishing distinctive textural references on its external façade while generating different qualities of spatial volumes inside. Photos on pp. 62–75 by Kolitha Perera.

第 62-63 页：图书馆围绕着现有的景观元素蜿蜒而行，创建内外空间，架构景观，并与场地建立几何关系。对页：蝴蝶状的屋顶随丘而下，在外立面上留下独特的纹理，在内部形成了各种体积的空间。

Opposite: Compared to the rammed-earth walls that rest firmly on the land, the steel pilot is float above rocks, thereby making different physical and ecological relationships with the site. This page, above: The building wraps around an internal courtyard which expands and integrates with the existing site morphology. This page, below: View from the staircase leading up to the research hub on the hill. The journey encounters with varied symbioses between the natural and the man-made.

对页：与稳固叠立在土地上的夯土墙不同，钢柱漂浮于岩石上方，从而与场地形成了不同的物理和生态关系。本页，上：建筑环绕着与场地融为一体的中庭；本页，下：直达山上研究中心的楼梯。在旅途中，可以邂逅自然与人为的多样共生。

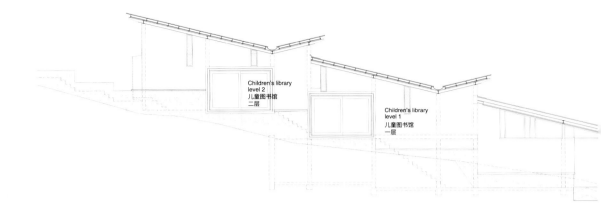

North elevation of building B and A (scale: 1/200) ／ B 建筑和 A 建筑北立面图（比例：1/200）

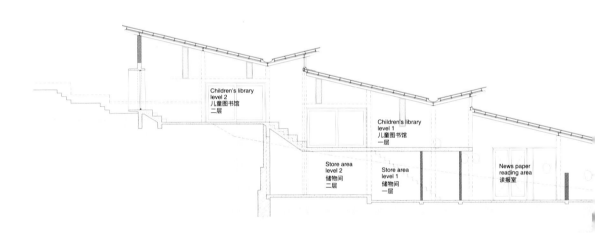

Section X of building B and A (scale: 1/200) ／ B 建筑和 A 建筑 X 剖面图（比例：1/200）

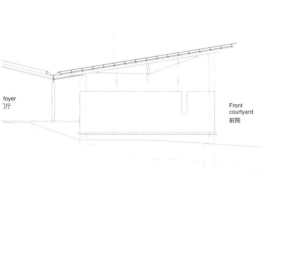

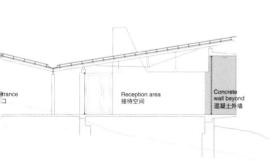

The library project, built by soldiers with the assistance from its local community, focuses on the building process as much as the building as a physical artifact, to celebrate a specific understanding of sustainable architecture derived from the very structure of its making. Impinging on its vocational training strategies and environmental planning, the project attempts to heal social wounds, build workforce capacity, disseminate knowledge, appreciate sustainable building, and strengthen social relations. The single-storey building spans nonchalantly across the landscape, resting on soil through rammed-earth walls and floating over rocks held aloft by galvanized iron tubes. The building informally wraps around a courtyard which is an extension of the external landscape. Its placement on site, made to accommodate existing trees, follows the scale of adjacent buildings and acknowledges the natural life of its physical setting. A series of formal and informal platforms for reading is organised in and around the building; its spatial progression unfolds as an experiential journey across diverse volumes, framed views, and blurred definitions between inside and outside. To overcome the general deskilling of construction workforce in the country, and to promote the army's participation in post-war reconstruction, the project explored the possibility of using real building projects as training grounds for skill development. With early preparations, training tasks were built into the design – this approach could be extended across the building industry as a policy to build workforce capacity.

Text by Milinda Pathiraja,
edited by Channa Daswatte

Credits and Data
Project title: Ambepussa Community Library
Client: Sinha regiment, Sri Lanka Army
Location: Sri Lanka Army, Ambepussa, Sri Lanka
Design: 2013
Completion: 2015
Architect: Robust Architecture Workshop
Design Team: Milinda Pathiraja, Ganga Rathnayake, Upul Piyananda, Kolitha Perera
Project Team: A. Wickramasinghe (structural engineer);
    Sinha regiment, Ambepussa (mechanical engineer / construction);
    K.R. Prematillake, Nimal Somasundara (lead training supervisors)
Project area: 1,402 m^2
Project estimate: USD 170,767 (31 million Sri Lankan Rupees)

This page: View of the main reading room. The projection of the linear spatial volume towards the landscape frames the views beyond, while facilitating cross-ventilation and shading.

本页：主阅览室。突出的线性空间裁切了远方的景观，同时也有利于通风和遮阳。

这个图书馆项目由军队士兵在当地社区的协助下建造，旨在传达对基于自身结构的可持续建筑的独特理解。其焦点不仅在于建筑本身，还注重其建造过程。该建筑影响了职业培训的策略和环境规划，试图治愈社会创伤、培养员工能力、传播知识、推广可持续建筑，并加强社会关系。这座单层建筑的夯土墙置于大地，由镀锌铁管支撑从而浮于岩石之上，从容地越过景观。建筑围合出不规则的中庭，作为外部景观的延伸。建筑布局是为了适应场地现存的树木，同时呼应相邻建筑物的尺度，以空间布局来尊重、回应自然。图书馆内外配置了各种阅读空间。建筑随着空间的展开徐徐展现，带来了跨越体积、框架视野以及内外界线的体验之旅。为了弥补斯里兰卡建筑工人普遍技术不足的缺陷，鼓励军队参与战后重建，该项目探索了将实际建设项目用作技能发展培训的可能性。这种将预先准备好的培训任务内置于设计中的方法可以扩展到整个建筑行业，作为提高建筑工人能力的一项政策。

米林达·帕蒂拉贾 / 文
钱纳·达斯瓦特 / 编

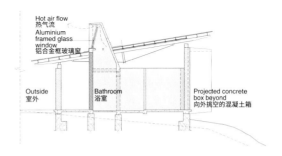

Section W of building A ／ A 建筑 W 剖面图

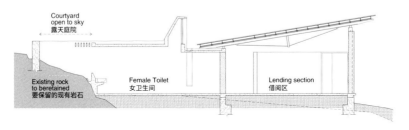

Section Z of building A (scale: 1/200) ／ A 建筑 Z 剖面图（比例：1/200）

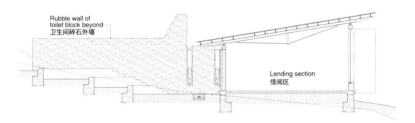

Section Y of building A ／ A 建筑 Y 剖面图

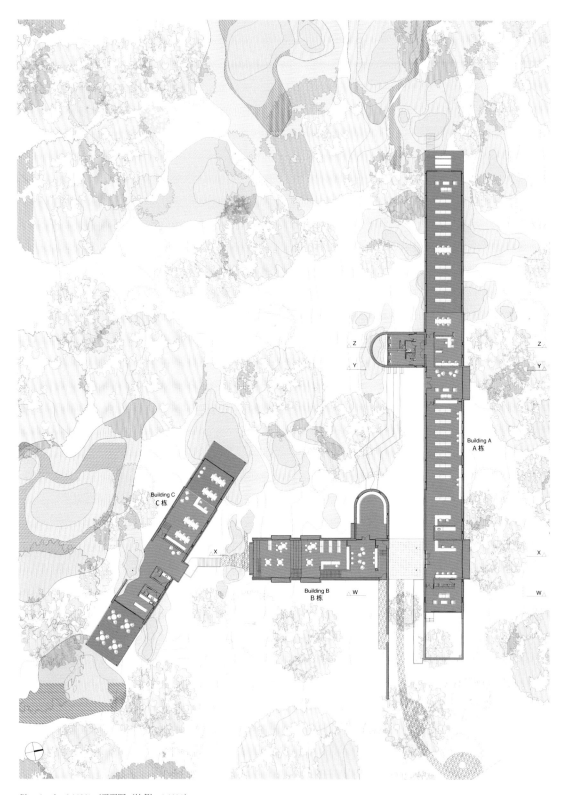

Plan (scale: 1/600) ／平面图（比例：1/600）

pp.72-73: The building is articulated along an experiential route of framed views, vistas and panoramas – synchronizing a variety experiences between the inside and outside. This page: View of the children's library. The play of colour, texture, light, volumes and views entices the young users to use these spaces freely.

第 72-73 页：沿窗景、风景及全景的路径连接起了该建筑，同时营造出丰富的内外空间体验。本页：儿童图书馆。通过色彩、材质、光影、体量和视野的使用设计区分，鼓励小孩子们自由使用。

Philip Weeraratne
Kotte Residence
Sri Jayawardenapura Kotte 2011–2014

菲利普·韦拉拉特尼
科特公寓
科特 2011–2014

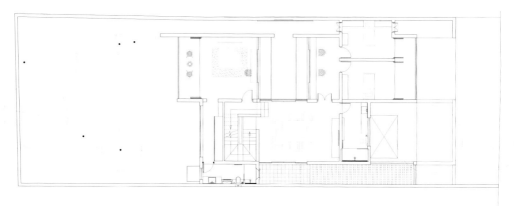

First floor plan ／二层平面图

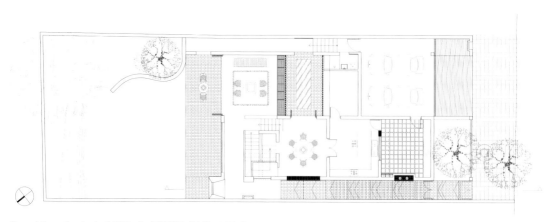

Ground floor plan (scale: 1/300) ／一层平面图（比例：1/300）

Built for a young family within the confines of a restricted urban plot, the design was primarily influenced by the linear nature of the site. A long double height walkway with a skylight on top is used as a link between the inside and outside. The living room opens out towards the garden and the old "beli" (Aegle marmelos) tree that was preserved carefully during construction. A granite stone lined wall retains the earth around the tree.

The simplicity of spaces is offset by the textural richness of the materials used; the local teak timber windows contrasted with the old cobblestones, the rough plastered walls contrasted with the cut and polished cement floors. The reflection of the sky can be enjoyed from the dining room through a shallow reflecting pond. The green lawn and creepers on the walls provide gentle backdrops which can be appreciated from both within and outside of the house. Beauty is perceived from the balance between elements and textures with the play of light that changes throughout the day. The arrangement and progression of spaces lend itself to a simple and modern experience of tropical living.

Text by Philip Weeraratne

Credits and Data
Project title: Kotte Residence
Client: Dinesh and Bimali De Mel
Location: Sri Jayawardenapura Kotte, Sri Lanka
Design: January 2011
Completion: January 2014
Architect: PWA Architects
Design Team: Philip Weeraratne (principal), Sumith Perera, Gayan Karunarathne
Project Team: STEMS Consultants Pvt Ltd (structural engineer); Nimal Perera Associates (services engineer); DV Construction Pvt Ltd (contractor); Jacob Pringiers (furniture)
Project area: 363 m^2 (GFA)
Project estimate: USD 171,100

p. 77: Elevation view of the house. The dialog between in and out is modulated through the treatment of openings, with different sizes and materials being used. Opposite, above: A play of levels, as well as elements like sunlight and water, adds to the atmospheric quality between spaces. Opposite below: The living room is limited to a muted palette so that the vibrant tones from the outside are brought in. Photos on pp. 76–81 by Waruna Gomis.

第77页：建筑外立面。不同尺寸和材料的入口营造出内外之间的对话。对页，上：对高差以及诸如阳光、水之类元素的运用增加了空间里独特的氛围；对页，下：客厅色调以柔和为主，以便外面充满活力的色彩投射进来。

这座建筑坐落在城市街区，住户是一个年轻家庭，建筑的设计主要受场地线性特点的影响和限制。狭长带状天窗的双层步行道连接室内外。客厅通向花园和建造过程中被精心保护下来的古老"贝利"（木橘）树。衬有花岗岩的石壁留住了树木周围的泥土。

简洁的空间与丰富的材料互补。当地柚木制成的窗户与陈旧的鹅卵石、粗糙的抹灰墙与切割抛光的水泥地板形成对比。在餐厅里可以欣赏倒映在浅池中的天空。墙上的绿色草坪与爬山虎成为温柔的背景，无论在屋内屋外，都可以观赏。随着一日之中光线的变幻，建筑元素和纹理之间的平衡展现出美感。空间的布置与变化为热带生活创造了简单而现代的体验。

菲利普·韦拉拉特尼／文

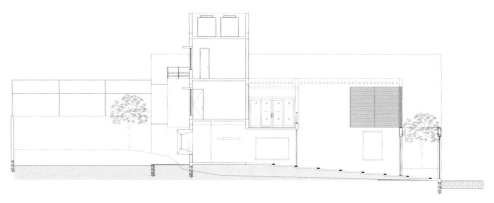

Section (scale: 1/300)／剖面图（比例：1/300）

Opposite: View along the weathered cobblestone walkway leads to a door that is made out of repurposed timber. Light filtering through pergolas gently washes over the wall and floor.

对页：风化的鹅卵石小路。道路通往一扇由旧木制成的门。光穿过藤架，轻轻地洒在墙壁和地板上。

Robust Architecture Workshop
House 412
Pannipitiya, Colombo 2011–2014

罗伯斯特建筑工作室
住宅 412
科伦坡，潘尼皮蒂亚 2011–2014

This 4 bedroom residential design has been formally conceived of as a "wall" expanded along its Cartesian coordinates, an object of 3-dimensional complexity, not just a planar element. Moving away from the typical anti-urban stance of placing a house inside a tall perimeter wall – common to most Sri Lankan suburban houses, this design places the main structural elements perpendicular to the street (as seen from plan), thereby immediately opening the street façade at the pedestrian level.

Upon entering the house, the threshold is prolonged; many episodes are encountered in a sequence as the "wall" incorporates window seats, daybeds and a verandah. They provide incidental connections and oblique views to the surrounding neighbourhood, while broadening the zone of the wall to function as a modulator of light and a device for weather protection along its periphery.

These episodes lead to the heart of the house, a large space that embraces the dining, kitchen and living areas. Along the way, one is encouraged to move through these spatial volumes of diverse widths and heights, thereby generating a poetic journey threading across horizontal and vertical spatial connections.

Text by Milinda Pathiraja, edited by Channa Daswatte

Credits and Data
Project title: House 412
Client: Mr. and Mrs. E.A. Ponnamperuma
Location: Pannipitiya, Colombo, Sri Lanka
Design: 2011
Completion: 2014
Architect: Robust Architecture Workshop, Pulina Ponnamperuma
Design Team: Milinda Pathiraja, Pulina Ponnamperuma, Ganga Rathnayake
Project Team: J.H.T. Chandrarathne (structural engineer), Prabath Construction Enterprise (construction and mechanical engineer)
Project area: 304 m²
Project estimate: USD 74,366 (13.5 million Sri Lankan Rupees)

Ground floor plan (scale: 1/300) ／一层平面图（比例：1/300）　　*Upper floor plan* ／上层平面图

p. 83: General view of the house. The architecture attempts to respond to the spatial rhythms and morphological flows of the suburban landscape it is part of. Opposite, above: The articulated building mass opens out to the street, thus becoming an urban incidence for the curiosity of anyone who passes by. Opposite, below: Entrance to the house. The experience starts with moving through solid building masses before opening out to a sequence of internal spatial volumes. All photos on pp. 82–91 by Kolitha Perera.

第 83 页：住宅全景。建筑试图回应所处郊区景观的空间节奏和形态流动。对页，上：建筑整体向街道敞开，作为一个城市要素引起了路人的好奇心；对页，下：住宅入口。建筑体验始于通过厚重的体量进入一系列室内空间。

这个四居室住宅在形式上是直角坐标上扩展的"墙",它不仅仅是一个平面元素,还是错综复杂的三维集合的组成部分。与大部分斯里兰卡郊区的房屋截然不同,它放弃了在高墙之内安置住宅(这是一种拒绝都市的姿态),而是将主要结构放置在与街道垂直的位置(从平面图上可以看到),从而使房屋立面直接向步行街道展开。

当进入住宅,入口向屋内延伸。沿着"墙",可以依次进入各种空间,诸如靠窗的座位、沙发床和阳台。

它们提供了与周围居民区意想不到的联系和倾斜视野,同时拓宽墙壁的领域以充当光的调制器和抵御气候的外围装置。这一系列元素都通向此建筑的心脏,一个兼具餐厅、厨房和起居功能的宽敞空间。穿过这些高低宽窄各异的房间,一段穿越水平和垂直空间联系的诗意旅程就会由此开启。

米林达·帕蒂拉贾 / 文
钱纳·达斯瓦特 / 编

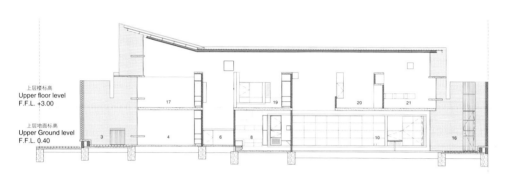

Section A (scale: 1/300) / A 剖面图(比例:1/300)

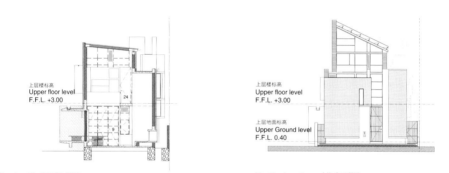

Section B / B 剖面图   North elevation / 北立面图

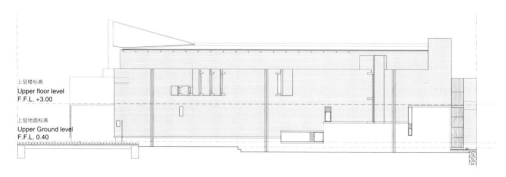

West elevation (scale: 1/300) / 西立面图(比例:1/300)

p. 87: Interior view facing the entrance of the house. The large openings along the corridors visually connects the interior living spaces with the life outside. pp. 88–89: The internal spaces open out onto the garden through a permeable envelope framed by overhanging building elements. Opposite: View from the living space. Voids are carved out to create visual connectivity at different levels. This page: Interior view from the kitchen that opens out onto the garden.

第 87 页：从建筑内部看向入口。视觉上，沿着走廊的大开口将内部起居空间与室外生活联系在一起。第 88-89 页：内部空间通过一个悬挑的围护结构通向花园。对页：起居空间。为创建不同层次的视觉联系雕刻出挑空空间。本页：从厨房看向花园。

Palinda Kannangara
Studio Dwelling at Rajagiriya
Buthgamuwa Road, Rajagiriya 2013–2015

帕林达·堪纳卡拉
拉贾吉里亚的建筑师寓所
拉贾吉里亚,布加穆瓦路　2013–2015

This is an office and residence of an architect, located by a marsh, in Rajagiriya Sri Lanka. Although located along urban fringe near a series of high-rise buildings, and close to the main road, the building is designed like a fortification. It is sealed from the Colombo heat (with specially designed double screens to limit western and southern exposure), traffic and noise of the road; but once within reveals unexpected views of the adjoining marsh and is totally permeable to the natural setting. Located on a small footprint, the building comprises of three levels and a roof terrace – the ground floor consists of parking, kitchen, model-making room and a guest suite, with each room opening onto a courtyard.

The first floor houses the lobby and work space. The meeting area, lounge, library, including a northern wing comprising of a bedroom with a balcony, and an open-to-sky bathroom, are located on the second floor. The upper most floor (roof terrace) opens up to the rooftop gardens with a living and entertainment pavilion. The building plays with volumes to create a variety of areas for living, work and leisure; as well as with materials and tectonic devices to create a cooler microclimate within the building. It provides an example for living and working within an urban realm, while still negotiating issues of the public and private within the same building.

Text by Palinda Kannangara,
edited by Channa Daswatte

Credits and Data
Project title: Studio Dwelling at Rajagiriya
Location: 316 Lake Side, Buthgamuwa Road, Rajagiriya, Sri Lanka
Design: 2013
Completion: 2015
Architect: Palinda Kannangara
Design Team: Palinda Kannangara, Savindrie Nanayakkara
Project Team: Varna Shashidhar (landscape architect);
   Ranjith Wijegunasekara (structural engineer)
Project area: 270 m^2 (GFA), 370 m^2 (built area), 450 m^2 (site area)
Project estimate: USD 162,380 (24 million Sri Lanka Rupees)

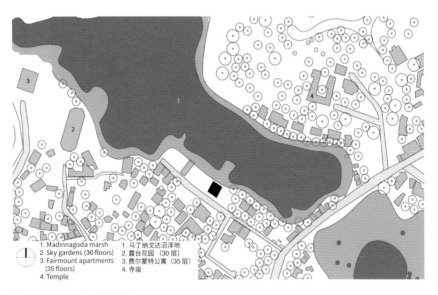

Site context map ／基地环境图

pp. 92-93: The central parking space, paved with reclaimed stone pavers from demolished tea estate, provides a framed view of the marsh. Photos on pp. 92-93, 96-97, 101, 102 below by Mahesh Mendis. Opposite: Elevational view of the double-layered wall. With an air gap in-between the two walls, the outer wall is made of un-plastered bricks with voids, while the inner wall is made of concrete. Photos on pp. 95, 99, 100, 102 above by Sebastian Posingis.

第 92-93 页：中心停车场，铺筑停车场的再生石材来自于被拆除的茶园。透过停车场，可以看到沼泽地。对页：双层构造墙的立面。两墙之间有空隙，外墙由未抹灰泥的砖砌成并留有空隙，内墙则由混凝土制成。

1. Parking/extension of rain garden	10. Entrance court	19. Conference area
2. Staff kitchen	11. Entrance lobby	20. Library
3. Staff toilet	12. Lounge	21. Living/ Entertainment pavilion
4. Guest suite	13. Office	22. Pantry/Dining
5. Store room	14. Panel room/stores	23. Bathroom
6. Model room	15. Fish pond	24. Shallow runner (Fish Pond)
7. Courtyard	16. Prayer niche	25. Herb garden
8. Rain garden	17. Bedroom	26. Rooftop garden
9. Spiral staircase	18. Bathroom	27. Paddy

1. 停车场／雨水花园扩建	10. 入口庭院	19. 会客厅
2. 员工厨房	11. 入口大堂	20. 图书馆
3. 员工厕所	12. 休息室	21. 起居／娱乐亭
4. 客房	13. 办公室	22. 食品储藏间／餐厅
5. 储藏室	14. 配电室／仓库	23. 浴室
6. 模型制作室	15. 鱼池	24. 浅槽（鱼池）
7. 中庭	16. 圣龛	25. 草本园
8. 雨水花园	17. 卧室	26. 屋顶花园
9. 螺旋楼梯	18. 浴室	27. 稻田

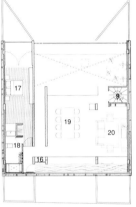

Second floor plan ／三层平面图

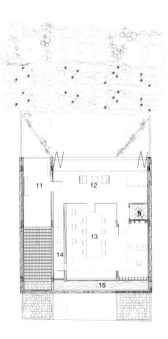

Roof floor plan ／屋顶平面图

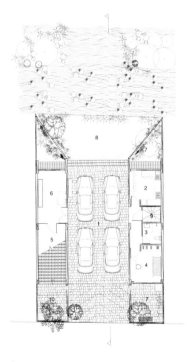

Ground floor plan (scale: 1/250) ／一层平面图（比例：1/250）

First floor plan ／二层平面图

pp. 96-97: The double height lounge area and office overlook the panoramic marsh landscape through specially designed 4.6 meters high windows. p. 99, above: The dramatic play of light on the second floor from the air gap between the two walls. p. 99, below: View from the meeting and library area on the second floor. The breathing brick wall creates a filigreed light effect while providing continuous air circulation. Opposite: The specially designed vertical louver system located on the second floor can be operated to control the circulation of air entering from the gap between the wall layers. This page: From the ground floor, the visitor ascends and starts a journey through wide stone paved steps (made using reclaimed stones) which is reminiscent of ancient temples.

第 96-97 页：双层高的休息区和办公室，透过特别设计的 4.6 米高的窗户可以俯瞰沼泽全景。第 99 页，上：二层的光影变化由双层墙之间的空隙产生；第 99 页，下：二层会议和图书馆区。可呼吸的砖墙持续提供空气流通，同时还营造出浮雕般的光影效果。对页：二层设置了经过特殊设计的垂直百叶窗系统，用于控制从墙的间隙带来的空气循环。本页：访客将从底层进入，踏上宽阔的石铺台阶（用再生石头制成）开启体验旅途。这使人联想到古代寺庙。

This page, above: The rooftop pavilion overlooks an intensive green roof filled with biological ponds, seasonal paddy field and edible gardens. This page, below: The interior of the office uses dark concrete to prevent glare from the sunlight coming in.

本页，上：从屋顶凉亭可以俯瞰到绿色屋顶，其上布满了池塘、季节性稻田和菜园；本页，下：办公室内部使用深色混凝土，防止阳光直射刺眼。

这个建筑师的办公室和住宅依沼泽而建，位于斯里兰卡拉贾吉里亚。这座建筑在一系列高层建筑附近，尽管地处城市边缘，却如同一个沿着主路的堡垒。它阻隔了科伦坡热（特别设计的双屏限制了来自西部和南部的暴晒）和道路交通的噪声。一旦进入建筑，便可见到意想不到的沼泽景致，建筑可以完全融入自然环境。该建筑占地面积不大，由三层楼和一个屋顶露台组成。底层有停车场、厨房、模型制作室和客房，每个房间都面向中庭。二层设有大厅和工作空间。三层则有会议区、休息室和图书馆，其北翼还包括一间带阳台的卧室和一个开放的露天浴室。最顶层（屋顶露台）通向带有凉亭，可供休憩和娱乐的屋顶花园。该建筑利用不同体量创造出起居、工作和休闲娱乐等各种空间。对于材料和构造设备也使用了同样的手段，在建筑内创造出凉爽的微气候。这座建筑提供了一个在城市内生活和工作的范例，同时使用者仍可以在同一建筑内协商公共与私人问题。

帕林达·堪纳卡拉 / 文
钱纳·达斯瓦特 / 编

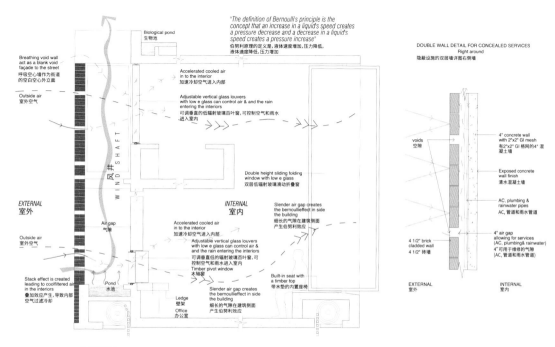

Detail drawing ／细部详图

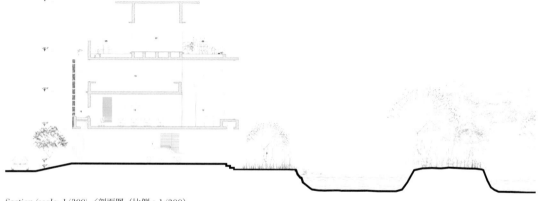

Section (scale: 1/300) ／剖面图（比例：1/300）

Chapter 2:
Conservation and Critical Reuse

第二章：
保护和批判性再利用

For many nations throughout the world, the boundary that defines the culturally significant past has become precarious because of rapid development and increased demand for urban land. In Sri Lanka today, the celebrated historic centers of its principal cities as well as smaller towns, villages and monuments are constantly under threat of destruction. Recent incidents have included the illegal overnight demolition of the Castle Hotel in Colombo that occurred in spite of protests by citizens. Even with its World Heritage City designation, Galle and its Fort, among its 18th and 19th century buildings, are not immune to the effects of greed, of individualism, of external economic drivers associated with development pressures. Such unmitigated actions are increasingly familiar yet bring ideas of preservation and reuse to the fore especially when linked to tourism and its deleterious effects. The historic need not only signify the past but also provide judicious approaches for thinking about the future.

For the following projects, close observation of the South Asian city's character and its associated buildings and spaces speak to processes that allow for a critical reorientation of how and where to build. Historical structures illustrate spatial patterns that cannot be replicated when they are destroyed. Interrogating the value of the old in concert with new strategies allows one to understand humane scales of careful construction, integration and mediation.

Sean Anderson

对于世界上许多国家而言，由于极速发展以及对城市用地需求的不断增加，具有重大文化意义的历史边界变得岌岌可危。在今天的斯里兰卡，主要城市的著名历史中心以及较小的城镇、村庄和古迹一直处于被破坏的危险之中。例如，最近不顾公民抗议，科伦坡的城堡酒店被连夜非法拆除。即使是有着18世纪和19世纪建筑物以及著名古堡的世界遗产城市加勒，也未能从贪婪的个人主义以及发展压力带来的外部经济驱动中幸免于难。这种彻底的破坏行为越来越为人们熟知，并将建筑遗产保护和再利用的思想带到台前，特别是考虑到旅游业及其带来的危害。对于历史的需求不仅指向过去，它也能提供思考未来的明智方法。

本节收录的项目密切观察了南亚城市的特征及其相关的建筑与空间，带来了重新定位如何以及在何处建设的批判性谈论。若历史结构被破坏，其描绘出的空间格局则将一去不复返。审视旧的价值，并使其与新策略相呼应，能够使人们理解人道尺度的精心构建、整合和调解。

肖恩·安德森

Pulasthi Wijekoon, Guruge Ruwani, Thusara Wadyasekara
Office Building for Colombo Municipal Council
Dr C.W.W Kannangara Mawatha, Colombo 2013–2019

普拉斯丁·维耶空，古鲁格·鲁瓦尼，苏萨拉·威迪亚塞卡拉
科伦坡市议会办公楼
科伦坡，C.W.W 坎纳卡拉·玛瓦莎博士路 2013–2019

Colombo Municipal Council (CMC) is envisaged as a government administrative office with a prominent public interface. From pensioners to taxpayers, architects, urban planners, government officials, and politicians are among the people who frequent the verandahs of CMC, making it a busy and dynamic environment.

The approach to designing a new office building for CMC in a double triangular site, located behind an old colonial building that house the council separated only by a through-site link, has been a sensitive and integrative one. Along with its strong programmatic inclination, the geometries of the site, the significant presence of the old CMC building, the park and the through-site link informs the design in more than one way.

The dynamics of a triangular footprint that closely follows the site, the stepping profile that carves out a verandah, and the network of overhead bridges, immediately engage the context and its user in a conversation. The interiors flow into a colonnaded verandah that blends with its landscape, and transforms into a through-site link that runs along the full length of the building under a network of pedestrian bridges. The greenery of the park and the lawn is extended through soft and hard landscaping which provided much-needed shading. The stepping section of the building encourages public activity by providing a shaded space at the ground level.

The new CMC building positively alters the streetscape using a visual gesture, by carefully removing a piece of physical mass from the new to reveal a powerful framed view of the old – in respect and homage to the important colonial building. The building form and façade treatment not only align with its programme but also enhance its performance. The stepping façade provides shading with its overhangs and further shades the interior with lace-like perforated screens, thus enhancing the passive thermal performance of the building.

Text by Pulasthi Wijekoon
Guruge Ruwani
Thusara Waidyasekara

科伦坡市议会（CMC）被设想为具有卓越公共交互功能的政府行政办公室。经常拜访 CMC 廊台的公民包括退休人员、纳税人、建筑师、城市规划师、政府官员和政客，他们使这里变得繁忙而充满活力。

CMC 新的办公大楼位于一块双三角形场地，采用了灵活而综合性的建造方法。该建筑处于议会所在的旧殖民建筑后面（被一条穿过场地的路隔开）。此外，该建筑还拥有多种设计手法：强力的程序化倾向、场地的几何性、旧 CMC 大楼的显著存在感、公园和贯穿场地的路。

紧随场地的三角形边界，沿着露台雕刻出的阶梯轮廓以及交错的高架桥网络能立刻将访问者引入与文脉的对话之中。室内空间流向屋外的柱廊，外廊与景观融合，并沿着建筑转变成贯穿场地的通道，衔接了城市人行天桥网络。公园和草坪绿化通过软质与硬质的景观设计得以扩展，还提供了必要的遮蔽。建筑的台阶在地面层提供荫蔽空间，以此来鼓励公共活动。

新的 CMC 大楼通过视觉手法积极地改变街景，例如从新建筑中谨慎地去掉一小部分实体，以呈现旧楼强有力的框架景象，表达对重要历史建筑的敬意。建筑师对建筑形态和外墙的处理不仅在功能上保持一致，还提高了性能。阶梯式立面的突出部分提供了荫蔽，并通过其花边状的穿孔屏进一步遮蔽内部，从而增强了建筑的被动热性能。

普拉斯丁·维耶空，

古鲁格·鲁瓦尼，

苏萨拉·威迪亚塞卡拉 / 文

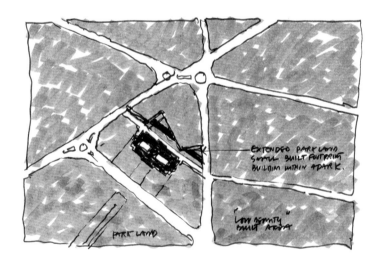

Sketch of the relationship between the new building extensions and sites
新建筑扩展和场地之间关系的手绘示意图

pp. 106-107: General view of the new extension building. A lace-like perforated screen is used to enhance the passive thermal performance of the building and create a unique identity for the building. pp. 108-109: A piece of physical mass is removed from the new building to reveal the important historical building behind. Photos on pp. 106-113 by Eresh Weerasuriya.

第 106-107 页：大楼扩展部分的全貌。花边状的镂空屏用于增强建筑的被动热性能，并使建筑具有了独特性。 第 108-109 页：新大楼中去掉了一块实体，以展示后面的重要历史建筑。

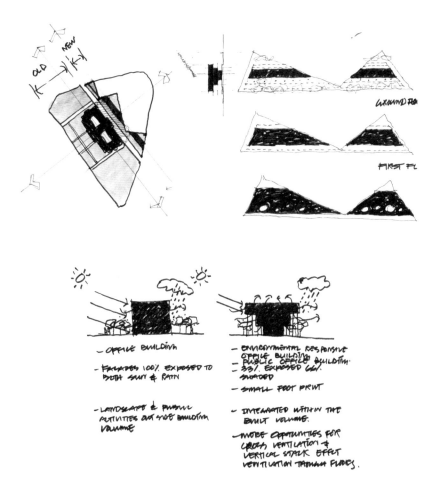

Study sketches ／研究手稿

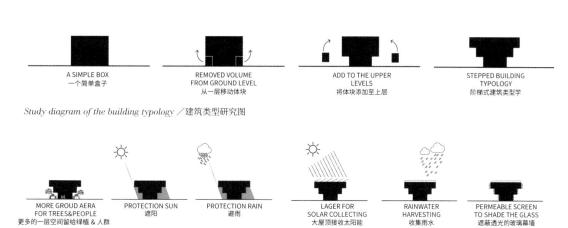

Study diagram of the building typology ／建筑类型研究图

Concept diagram ／概念分析图

Opposite: The through-site link, verandah and colonnade resonates with the old CMC building, providing much needed public space in the ground level. This page: A network of overhead pedestrian bridges provides additional connectivity between the two triangular buildings.

对页：外廊和柱廊是贯穿全场地的连接空间，与 CMC 旧大楼产生共鸣，也在地面层上提供了必要的公共空间。本页：高架人行天桥成为连接两个三角形建筑的通道。

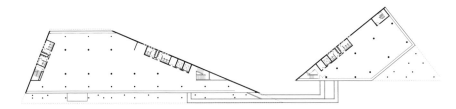

Second floor plan ／三层平面图

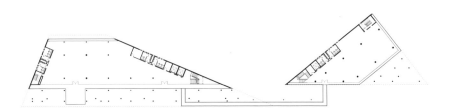

First floor plan ／二层平面图

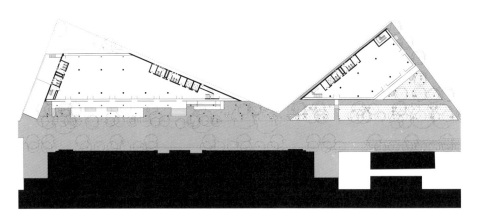

Ground floor plan (scale: 1/1200) ／一层平面图（比例：1/1200）

pp. 114-115: Perforated lace-like screen filters soft light into the interior spaces. Opposite: View of the main circulation space bathed in natural light. p. 119: General view of the open-plan office. Photos on pp. 114–119 by Prageeth Wimalarathne.

第 114–115 页：花边状镂空屏将柔和的光线过滤到内部空间中。对页：沐浴在自然光线下的主要交通空间。第 119 页：开放式办公室。

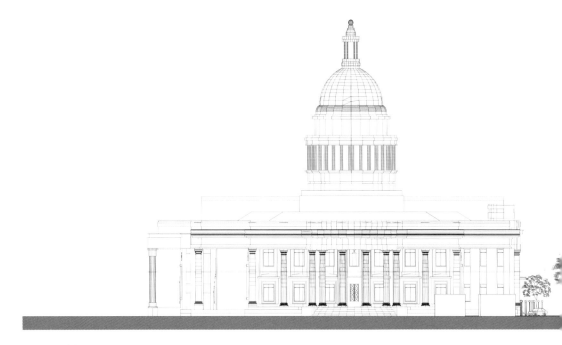

East elevation ／东立面图

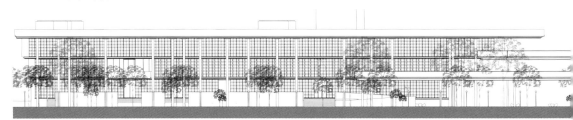

North elevation (scale: 1/500) ／北立面图（比例：1/500）

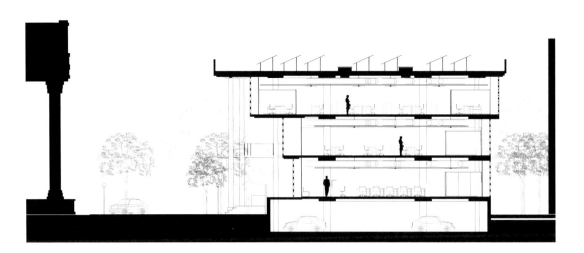

Short section (scale: 1/300) ／横向剖面图（比例：1/300）

Credits and Data

Project title: Office Building for Colombo Municipal Council
Client: Colombo Municipal Council
Location: C.W.W Kannangara Mawatha, Colombo 7, Sri Lanka
Design: 2013
Completion: 2019
Architect: Pulasthi Wijekoon, Guruge Ruwani, Thusara Waidyasekara
Design Team: Jagath Panawala, Nadeesh Senevirathna, Lakmal Galagoda
Project Team: Kowshika Gunasena (landscape architect); Prasad Disanayaka, Hasitha Gunasekara (quantity surveyor); Shiromal Fernando (structural engineer); N.R. Sooriarachchi, Jeffry Jayasingha (MEP engineer);
Rohitha Senanayaka (electrical engineer)
Project area: 8,334 m^2

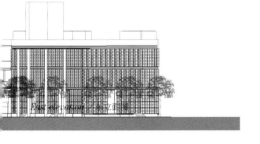

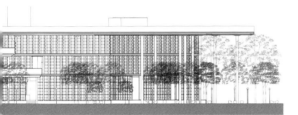

Thisara Thanapathy
TRACE Expert City
Maradana, Colombo 2013–2014

蒂萨拉·泰纳帕里
TRACE 专家城
科伦坡，马拉达纳 2013-2014

p. 120, above: Photo taken before renovation. Two story warehouse in its total abandoned state. p. 120, below: The 13 bays during the initial stage of renovation. p. 121: Series of open and closed spaces visually connected through the brick arches. Photos on pp. 120–121 courtesy of the architect. Opposite: View from the interior looking out into the green courtyard that connects the surrounding spaces together. Photo by Thilina Wijesiri. This page: Courtyards placed at alternating bays bring light and ventilation into the offices, integrate spaces with its greenery, and provide its inhabitants with places for informal activities. Photos on pp. 123–129 by Waruna Gomis

第 120 页，上：翻新之前的照片。两层仓库被完全废弃；第 120 页，下：翻新初始阶段的 13 个间隔。第 121 页：视觉上，一系列开放和封闭空间通过砖拱相连。对页：从室内看向中庭，绿色的庭院将周围空间连接在一起。本页：隔层中配置的庭院使办公室实现采光和通风，将空间与绿化结合起来，并为居民提供休闲活动场所。

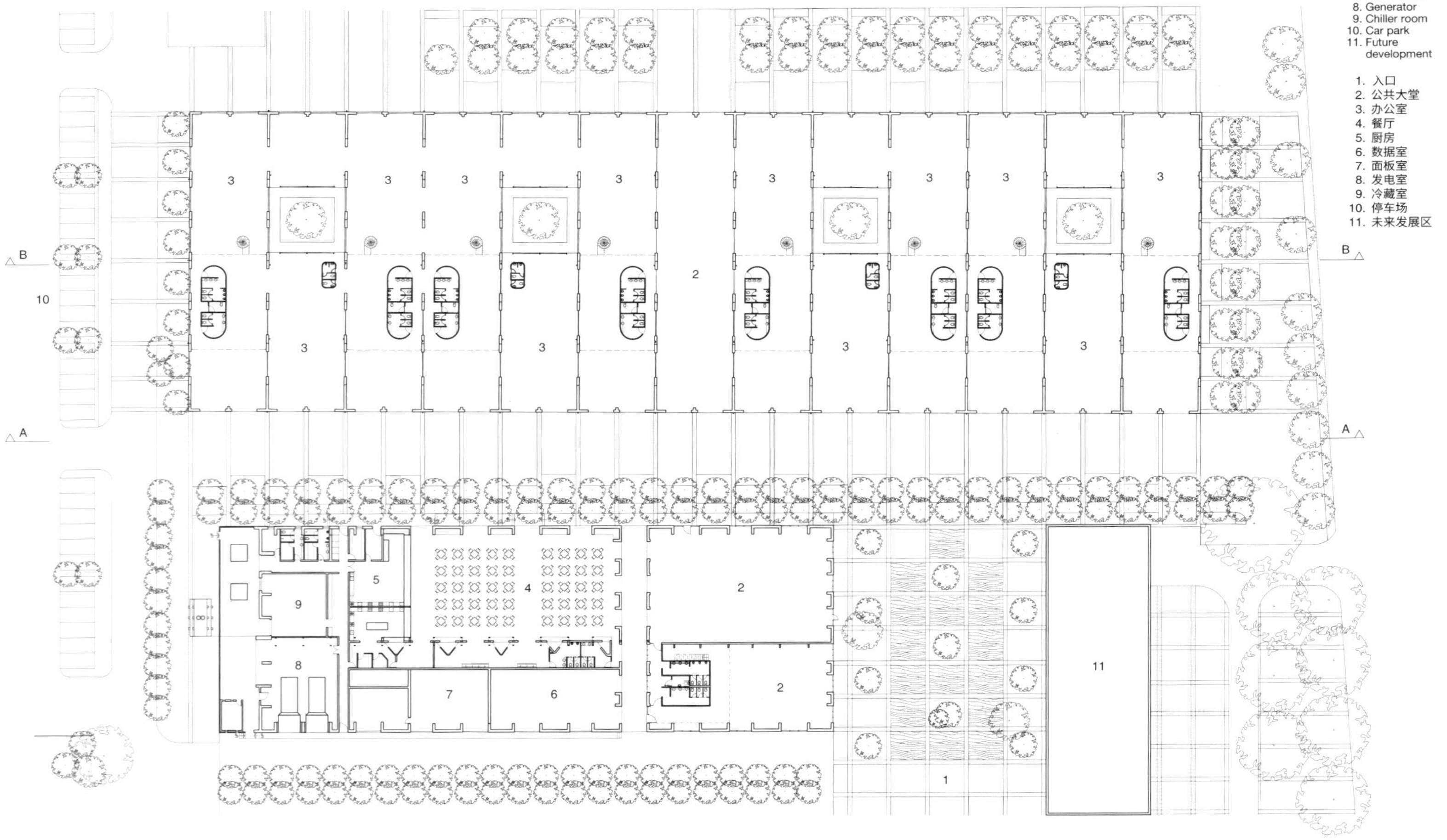

1. Entry
2. Common lobby
3. Office
4. Restaurant
5. Kitchen
6. Data room
7. Panel room
8. Generator
9. Chiller room
10. Car park
11. Future development

1. 入口
2. 公共大堂
3. 办公室
4. 餐厅
5. 厨房
6. 数据室
7. 面板室
8. 发电室
9. 冷藏室
10. 停车场
11. 未来发展区

Plan (scale: 1/1,000) / 平面图（比例：1/1,000）

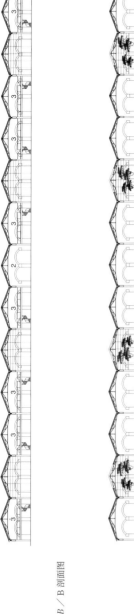

Section B / B 剖面图

Section A (scale: 1/1,000) / A 剖面图（比例：1/1,000）

The project, occupied by office spaces for a technological innovation park, was a part of an urban renewal program conducted by the government to revitalize historic buildings. A fast-track construction was proceeded by security forces which required quick design decisions and close consultation. An 18th century British warehouse complex was restored and reused to cater to the needs of an environmentally friendly, tropical city. By exposing the historical layers of the city, and reinforcing the green spaces in the open areas between them as shaded courtyards with additional planting and water bodies, the project attempts to improve the livability of this inner city area of old Colombo.

The historical industrial characters of the buildings are carefully preserved with sensibly combined, simple and graceful modern interventions. Retaining the exposed brick and steel trusses, the buildings were then covered with a zinc and aluminium corrugated roofing, with proper fenestration provided using clear glazing on steel or aluminum frames, along with exposed ducts, pipes and service elements. The proposed design celebrates the industrial character of this historic building. In the 13-bay building, a white mezzanine floor was introduced in alternative bays to increase usable space and as a complementary modern feature. This gives an illusion of a floating cantilever stretching from the small solid oval block, below which contains the toilets and other services. The alternative shaded courtyards within offer natural light and ventilation, connect spaces visually, and bringing greenery into the working spaces.

Text by Thisara Thanapathy,
edited by Channa Daswatte

Credits and Data
Project title: TRACE Expert City
Client: TRACE
Location: Maradana, Colombo, Sri Lanka
Design: 2013
Completion: 2014
Architect: Thisara Thanapathy Architects
Design Team: Thisara Thanapathy (principal), Denash Gunawadhane, Rafidh Rifaadh
Project Team: R.S.K. Thrimavithana (structural engineer); Kosala Kamburadeniya, Chandana Dalugoda (M&E engineer)
Project area: 12,077 m^2
Project estimate: USD 3,900,000 (700 million Sri Lankan Rupees)

TRACE专家城是政府为振兴历史建筑而进行的城市更新计划的一部分，场地内聚集着技术创新园区的办公区域。建设由治安部队快速推进，并要求迅速的设计决策和紧密的咨询相配合。该项目修复并重新启用了18世纪的英国仓库群，以满足环境友好热带城市的需求。设计试图通过展示城市的历史层次、增强建筑间开放的绿色空间，包括额外植被和水域的荫蔽中庭，来提升科伦坡旧城区的宜居性。

该建筑的历史工业特征通过合理的组合、简单而优美的现代干预措施得以精心保存。裸露的砖和钢桁架被保留，并使用锌和铝的波纹屋顶覆盖其上，还合理设置了带钢或铝框的透明玻璃窗，配以裸露的导管、管道和服务元素。这样的设计方案凸显了这一历史建筑的工业特色。在这座有13个间隔的建筑之间插入白色隔层以增加可用空间，作为现代功能的补充。隔层给人一种从小块椭圆实体伸展出漂浮悬臂的错觉，其下是洗手间和其他设施。室内的荫蔽中庭提供了自然采光和通风，在视觉上将工作空间相连接，同时引入绿色景观。

蒂萨拉·泰纳帕里／文
钱纳·达斯瓦特／编

pp. 126–127: View of the open public space between the colonial buildings. A row of trees directs one's focus onto the city's landmark – the lotus tower. Opposite, above: View of an informal meeting space. Opposite, below: The exposed brick surfaces and services highlight the utilitarian past of the structure as warehouses.

第126-127页：殖民时期建筑之间的开放公共空间。一排树木将人们的视线引导至城市地标——莲花塔上。对页，上：休闲聚会空间；对页，下：裸露的砖块表面和设施凸显了该建筑作为仓库功能的历史。

Dilum Adikari
Arcade Independence Square
Independence Ave., Colombo 2012–2014

迪卢姆·阿迪卡里
独立广场购物中心
科伦坡,独立大道 2012–2014

The complex of buildings was constructed in 1889 to serve as the Jawatta Lunatic Asylum in what was then considered a rural edge of the recently developed garden city of Cinnamon Gardens in Colombo. After the asylum was moved further out to the suburbs of Colombo, this complex was occupied by a series of government institutions including the Sri Lanka Broadcasting Corporation (SLBC), the Public Administration Department, the Auditor General's Department, and the Government Analyst's Department. Throughout this period, numerous additions were made to the complex to accommodate the various, often conflicting, needs of the different departments. The main entrance building was the focal point of Torrington Square, one of the garden squares in the city plan. Soon after independence in 1948, this square became an important focus of the city with the Independence Memorial Hall being built at its center. Later, this entrance building was converted into the offices of the chief minister of the western province, before it was once again being abandoned for new offices elsewhere. With its significant location, now renamed Independence Square, this area became one of the key focuses for urban renewal in 2009, after the end of the 30-year civil war in Sri Lanka. The urban fabric of this particular area had deteriorated significantly. Therefore, to revitalize the area, the Urban Development Authority prepared development proposals, declared, and gazetted Independence Square with its surroundings as a special development area. This complex of buildings, that was formally the Lunatic Asylum, became the focal project in this urban renewal scheme.

As for the buildings, the rich colonial architectural characteristics and dignity were lost due to its misapplication, lack of maintenance, and haphazard developments on both the inside and outside of the building complex. This had led to serious dilapidation in some areas of the building. However, a condition report obtained from National Building Research Organization confirmed of its structural stability. Due to the absence of drawings of these buildings, a measured drawing was carried out. New constructions that existed within the complex were demolished, wall plasters in a highly deteriorated condition were removed, arches in the arcades were filled up over time, and slowly, other aspects of the original design could be clearly identified. After removing all the elements added haphazardly overtime, a new reuse plan could be worked out. The simple plan could easily be adapted into a publically accessible shopping precinct. The main entrance was set to be from Independence square, through the clock tower wing, with its path leading to spacious promenades and the central section of the building extending across in an H-shape. The original design, consisting of wide corridors with arches, doors and windows were easily adapted from the old hospital function to its new use.

Today, the Independence Arcade is a symbolic public landmark in the city that celebrates its significant British colonial past.

Text by Channa Daswatte

---

pp. 130–131: General view. A water feature with a glass deck was introduced in the courtyard as a form of urban intervention, after the recently constructed building in the compound was demolished. Photos on pp. 130–141 by Prageeth Wimalarathne unless otherwise noted. Opposite, above: Photo taken during the demolition of the ad hoc constructions abutting the original building. After the removal of these constructions, an access from Independence Avenue is created. Opposite, below: Before, the windows of the original building were closed using masonry walls, as seen in the photo. After renovation, these openings are recreated and their windows, restored. Photos courtesy of the architect.

第 130-131 页：外观。作为城市干预的一种形式，在此建筑群中最近建造的楼屋被拆除后，庭院引入了带有玻璃甲板的水景。对页，上：拆除与原始建筑相邻的临时建筑时拍摄的照片。移除这些构造后，建造了一条通往独立大道的新通道；对页，下：如照片所示，之前的窗户被砖石墙封闭。装修后，这些开口被重建，窗户也得以修复。

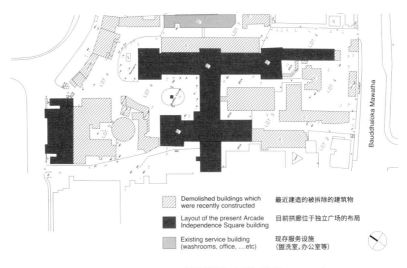

Site plan before demolition (scale: 1/2,500) ／拆除前的总平面图（比例：1/2,500）

独立广场购物中心建筑群最初是建于 1889 年的贾瓦塔精神病院。当时此地被认为是科伦坡刚刚发展起来的花园城市"肉桂花园"的边缘。在精神病院被进一步移至更远的科伦坡郊区后，这组建筑群被一系列政府机构占用，其中包括斯里兰卡广播公司（SLBC）、公共行政部、总审计部和政府分析部。在此期间，为满足不同部门的需求（通常是相互冲突的）而对建筑群进行了许多增建。主入口的建筑是托灵顿广场的焦点，该广场是城市规划中的花园广场之一。1948 年斯里兰卡独立后不久，就在广场中心修建了成为城市焦点的独立纪念馆。后来，这座入口大楼被改建为西部省省长办公室，之后又因为修建了其他新办公室而被弃用。此地因其重要的地理位置，现已更名为独立广场，是斯里兰卡长达 30 年的内战结束后，2009 年城市重建的重点之一。但这个区域的城市结构已大大恶化。因此，为振兴该地区，城市发展局策划了开发方案，并发出公告将独立广场与其周边地区列为特别开发区。就这样，曾经是精神病院的建筑群成了这个城市更新计划的重点项目。

至于建筑本身，由于其使用不当、缺乏维护以及建筑内外毫无规划的扩展，已经丧失了殖民建筑丰富的特色和尊严。这导致了建筑里某些区域的严重破败。好在国家建筑研究组织获得的状态报告证实了这里结构稳定。由于这些建筑没有专业图纸，因此还对其进行了测绘。增建部分被拆除，高度恶化的墙壁灰泥被移除，拱廊中的拱门慢慢地被填满。随着时间的流逝，原始设计的其他方面也可以被清晰地识别出来。在移除了所有随意增加的元素之后，便有了制定重建方案的可能。该建筑群简单的平面可以很容易地转换为面向大众的购物区。主要出入口被设在独立广场，穿过钟楼就可以来到宽广的步行道以及呈 H 形延伸的建筑中心。原始设计包括适用于旧医院功能的宽阔走廊，走廊还带有拱、门和窗，这些元素很容易被用于新设计。

今天，独立广场购物中心是该城市具有象征性的公共地标，代表着一段意义深刻的英国殖民历史。

钱纳·达斯瓦特 / 文

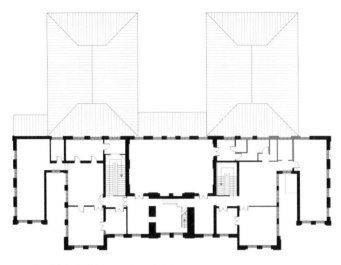

■ Demolished walls which were recently constructed
　近期被拆除的增建墙

Upper floor plan (part)／上层平面图（部分）

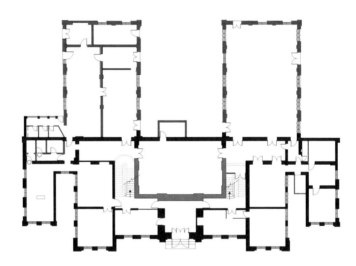

Ground floor plan (part, scale: 1/500)／一层平面图（部分，比例：1/500）

p. 135: A glazed roof is introduced to eliminate the gloominess of the lobby. Opposite: The closed arch is reopened and a cast iron handrail is used in replacement of the old timber handrail.

第 135 页：为改善昏暗的大堂，设置了玻璃屋顶。对页：恢复了闭合的拱门，并使用铸铁扶手代替老旧的木扶手。

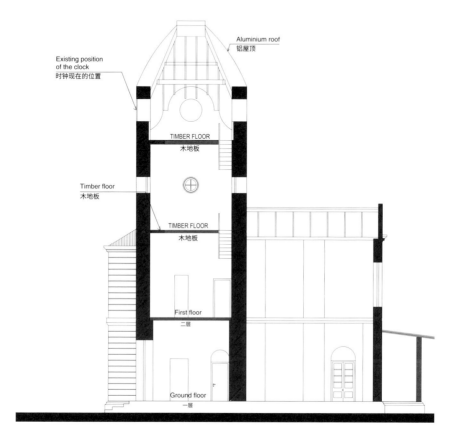

Section of west wing (part, scale: 1/200) / 西侧楼剖面图（部分，比例：1/200）

Opposite, above: View of the upper floor. The closed arches are now opened and a few artifacts from the old buildings are kept to eliminate the monotonous quality of the space. Opposite, below: View of the shops lined along the corridors. After the toilet blocks were removed, the spatial quality of the building improved.

对页，上：二层。曾封闭的拱门现已开放，并保留了旧建筑中的一些文物，以消解空间的单调性；对页，下：沿走廊排列的商店。拆除卫生间区域后，建筑的空间质量得到提升。

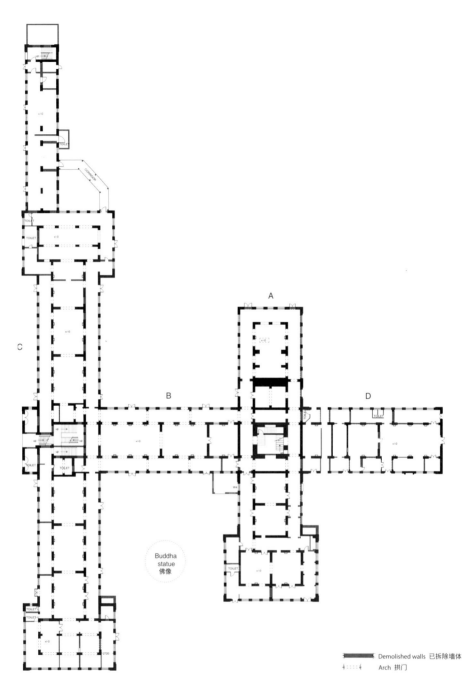

Ground floor plan of the main building (scale: 1/800) ／主建筑一层平面图（比例：1/800）

Credits and Data

Project title: Arcade Independence Square (Refurbishment of Former Auditor General's Department Building and Western Provincial Council Building)
Location: Independence Ave., Colombo, Sri Lanka
Design: July 2012
Completion: July 2014
Architect: Dilum Adhikari (Asst. Director, Urban Development Authority – UDA)
Design Team: directors, draughtsmen and technical officers of Project Management Division (UDA)
Project Team: Kosala Tennakoon (UDA, landscape architect); Capt. Hiran Balasooriya (Sri Lanka Navy, contractor)
Project area: 8,400 m^2
Project estimate: USD 5,233,170 (950 million Sri Lankan Rupees)

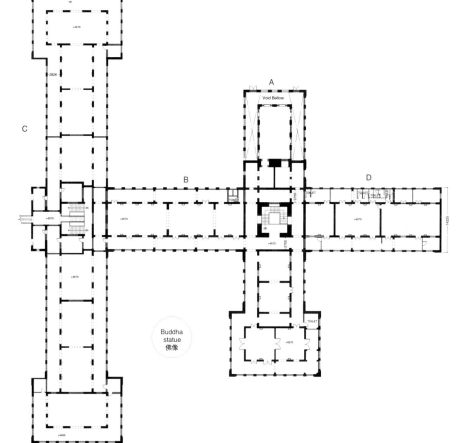

Upper floor plan of the main building ／主建筑上层平面图

Dilum Adikari
Dutch Hospital Shopping Precinct
Hospital Street, Colombo 2011

迪卢姆·阿迪卡里
荷兰医院旧址购物美食广场
科伦坡，医院路 2011

This building was constructed during the period of Dutch administration of the maritime provinces of Sri Lanka, from 1656 to 1798, in the heart of the city of Colombo – the area is commonly known as "The Fort". It originally belonged just within the ramparts on the southern edge of the fortified city. Though a definitive date is not available, a Dutch map drawn in 1732 shows the hospital on its present site; and a description by Christopher Schweitzer, a German who was in Sri Lanka from 1676 to 1682 in service of the Dutch, indicated that it has already been there in 1681.

This was the country's leading hospital during the Dutch occupation, and was primarily looking after the health of the officers and other staff serving under the Dutch East India Company. It occupies about half a hectare of land, which is a relatively large area in comparison to the size of the Fort. The building appears hardly to have changed during the last two centuries. With its large airy rooms and shaded verandahs, it served well for the needs of a hospital. Located close to the harbor, it was convenient to transport patients from the ships to the hospital.

With the advent of the British and the growth of the plantation economy, Colombo became a center of the newly industrialized economy. The British demolished the fortifications in 1880 as part of a city expansion scheme and the hospital was moved out – leaving behind only the building. Many of the key commercial institutions of the country were situated in the fort and it became a place with many of the grander British colonial buildings.

During the civil war in Sri Lankan, the fort area, which was also the center for the executives of Sri Lankan government and the residence of the president, was heavily fortified and many businesses moved out. By the end, in 2009, the government recognized the potential of the fort with its magnificent colonial building and it became a center of tourism for the city of Colombo. The conservation and renovation of the Dutch hospital precinct, previously used as a police station, was considered a key project in this development.

The process began with the demolition of the many ad hoc additions to the building that was not suitable to its original character – including an underground armor and toilet built in the rear courtyard. Eventually, however, all of these new constructions were removed except the underground armory; it was difficult to demolish due to its heavy structure. Now, it is used as an underground water sump.

After restoration, the rooms and spaces have wide corridors around the central courtyard, and the doors and windows were easily adapted into its new function after conservation – a shopping precinct with 13 shops, including restaurants, and is now a very popular destination for both its citizens and visitors in the old city.

Text by Channa Daswatte

p. 143: View towards Echelon Square through the arch at the main entrance. This page: General view from the front courtyard against a background of high-rise buildings at Colombo Fort. Photos on pp. 102–106 by Prageeth Wimalarathne.

第 143 页：从正门处的拱门看梯形广场。本页：以科伦坡城堡区的高层建筑为背景，从前院看到的全景。

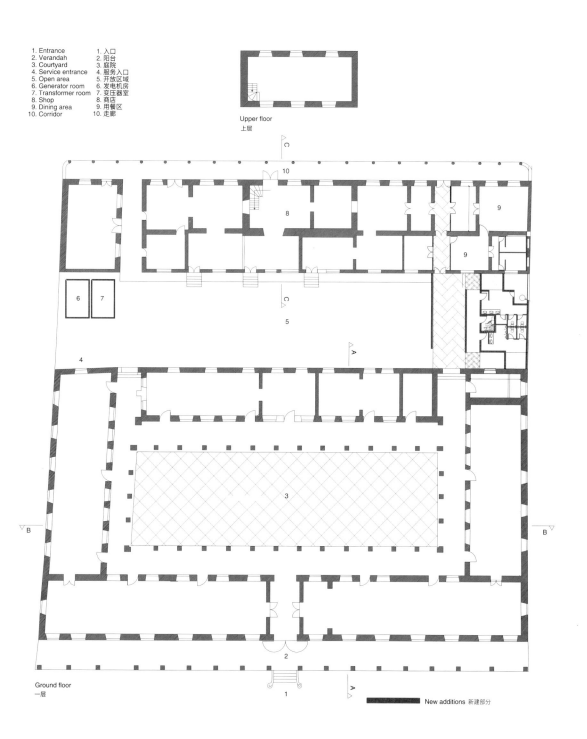

Plan, red lines indicate new additions to the building (scale: 1/500) ／平面图，红线表示新建部分（比例：1/500）

Credits and Data

Project title: Dutch Hospital Shopping Precinct (Renovation of Old Dutch Hospital Building)
Location: Hospital St, Colombo, Sri Lanka
Design: Feb 2011
Completion: Dec 2011
Architect: Dilum Adhikari (Asst. Director, Urban Development Authority – UDA)
Design Team: Champika de Silva (Director, Architect, UDA); draughtsmen and technical officers of Project Management Division (UDA)
Project Team: Lt. Col. E.P.H. Chandrarathne (iv ESR of Sri Lanka Army, contractor)
Project area: 3645 m²
Project estimate: USD 550,860 (100 million Sri Lankan Rupees)

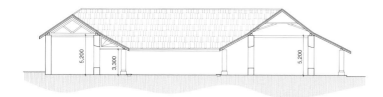

Section A / A 剖面图

Section B / B 剖面图

Section C (scale: 1/500) / C 剖面图（比例：1/500）

Opposite, above: View along the corridor of the front courtyard in the evening. Opposite, below: Photo of the rear courtyard at night.
对页，上：傍晚时分沿前院走廊眺望的景观；对页，下：后院的夜景。

Opposite: The existing window designs and materials were preserved from its original. This page, above: View towards Hospital Street. This page, below: View from the corridor leading to the rear courtyard.

对页：窗户设计和材料保留了原始状态。本页，上：从室内看向医院路；本页，下：从走廊看向后院。

这座建筑建于1656-1798年荷兰统治斯里兰卡的殖民时期，位于科伦坡的心脏地带"城堡区"。建筑最初坐落在要塞城南部边缘的壁垒内。尽管没有建设的确切日期，但1732年绘制的荷兰地图显示出该地有一家医院；而根据1676-1682年在斯里兰卡为荷兰人服务的德国人克里斯托弗·施韦泽的描述，1681年时建筑已经在那里了。

这是荷兰占领斯里兰卡期间最先进的医院，主要负责照顾荷兰东印度公司的管理层和其他工作人员。建筑占地约半公顷，与城堡区的整体规模相比，这是一个相对较大的区域。在过去的两个世纪中，这座建筑几乎没有变化。通风良好，房间宽敞，阴凉的阳台很好地满足了医院的需求。建筑位于港口附近，便于将患者通过轮船送到医院。

随着英国人的到来和种植园经济的发展，科伦坡成为新兴工业化经济的中心。作为城市扩张计划的一部分，英国人于1880年拆除了原有要塞，医院也被迁出，只留下了建筑。当时斯里兰卡许多主要的商业机构都位于城堡区之中，因此这里建有许多宏伟的英国殖民建筑。

在斯里兰卡内战期间，城堡区是斯里兰卡政府高官和总统官邸所在，设防严密，很多企业都被迁出。最后，在2009年，政府意识到了城堡区宏伟殖民建筑的潜力，使这里成了科伦坡的旅游中心。荷兰医院旧址在改造前是警察局，对其进行保护和翻新被认为是关键发展项目。

这个过程始于拆除建筑中那些将就的、不适用于原始特质的临时附加结构，包括在后院建造的地下军械库和洗手间。最终，除了结构笨重、很难拆除的地下军械库外，所有这些增建筑都被拆除。现在，地下军械库被用作地下水池。

整修后，中央庭院周围都设有宽阔的走廊，而门窗也根据新功能进行了改造。该建筑现在是一个拥有13家商铺（包括餐馆）的购物区。对老城区的市民和游客来说，这里是极受欢迎的目的地。

钱纳·达斯瓦特 / 文

Opposite, clockwise from top left: The demolition process of the ad hoc internal partition walls. Photo taken during the demolition of the toilet complex with a basement, located in the rear courtyard (at Spa Ceylon side). Access from Hospital Street was blocked before the renovation. Photo of toilet complex with a basement, located in the rear courtyard, during the process of demolition. The building was in a dilapidated condition. Photo of the building's interior before renovation. Photos on p. 107 courtesy of the architect.

对页，从左上顺时针方向：拆除临时内部隔墙的过程；在拆除后院（在锡兰水疗中心一侧）带有地下室的卫生间过程中拍摄的照片；在整修之前，无法从医院路进入建筑；在拆除过程中，位于后院的带有地下室的卫生间照片；该建筑处于残旧状态。整修前建筑室内的照片。

Channa Daswatte
Galle Fort Hotel
Galle Fort, Galle 2002–2004

钱纳·达斯瓦特
加勒古堡酒店
加勒，加勒古堡 2002–2004

The Galle Fort Hotel was rescued from what was a dilapidated wreck of a house that belonged to a significant merchant family from Galle. The original structure goes back to the 18th century with various additions made in the 19th and early 20th century as its wealth and public image called for. In the late 19th century when the port of Colombo was developed with the construction of a new breakwater, the family had moved their business to Colombo and the house in Galle was all but abandoned. The building was eventually used as a gem factory where rough gems were cut and made into sparkling baubles. The increased interest in the fort as a heritage precinct after it was declared a world heritage site in 1988 attracted various investors to purchase and renovate these old buildings as second homes or hotels. This is what drew its original owners back to the building.

Like many abandoned heritage buildings, a thorough clean up and preliminary removal of ad-hoc built walls and partitions revealed a magnificent house. The 18th century plan was representative of a Dutch colonial home of that period. The large verandah lead into a hallway flanked by two rooms, which in turn lead into the grand *Zaal* or hall. This opened onto another verandah that overlooked the rear courtyard and a magnificent double height colonnaded wing with low *piano nobile* of rooms on top, which may have replaced the original staff and kitchen areas. An investigation into its history recalled that these latest additions were done in the 1920s for a family wedding, including an eccentric archway between the entrance hall and the *Zaal*. Further cleaning revealed paintings from the 1920s on the bedroom walls which were carefully conserved and covered up.

The conservation of the building began by a clear recording of the existing fabric, and planning it based on the conservation rules set out by the Urban Development Authority and the Department of Archeology who supervised the process. The decision was made to add a new wing in a period style but different from the others, taking on the idea that the different periods are reflected in their different times. 16 rooms were carved out from the entire building complex, turning it into a small hotel for its guests to feel what it was like to live in an 18th century mansion when visiting the fort. Most parts of the interior design worked with local furniture and fabrics. In 2007, it was given a UNESCO Asia Pacific Heritage award for excellence in conservation and reuse.

Text by Channa Daswatte

Credits and Data
Project title: Galle Fort Hotel
Client: Karl Steinberg, Christopher Ong
Location: Church Street, Galle Fort, Galle, Sri Lanka
Design: 2002
Completion: 2004
Architect: MICD Associates
Design Team: Channa Daswatte, Dilrukshi Paranamanage
Project Team: Tissa Deerasekera (civil engineer); Asoka de Silva (contractor)
Project area: 1,500 m^2
Project estimate: USD 1,000,000,000

pp. 152–153: All the bedrooms in the now reconstituted hotel overlook the central courtyard which was originally the back garden of the merchant's mansion. The bedrooms now occupy both the original wing of rooms on the left and the new wing on the right. Opposite, above: View from along the new wing. During conservation, it was discovered that the house contains a variety of extensions from different periods. Opposite, below: View from the connection between the back verandah and the western suite block (original wing). Photos on pp. 152–161 by Dominic Samsoni.

第 152-153 页：这家经过改建的酒店中，所有卧室都可欣赏原来作为商人宅邸后花园的中庭。现在，卧室既占据了左侧旧馆的房间，又占了右侧的新馆。对页，上：从新馆看向中庭。在保护过程中发现住宅包含不同时期的加建；对页，下：背面阳台和西部套房区（旧馆）之间的通道。

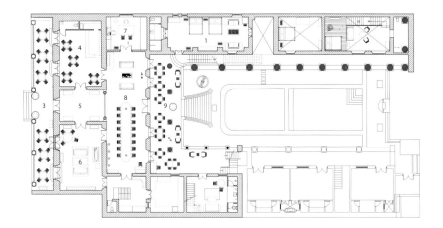

Entry level plan ／入口层平面图

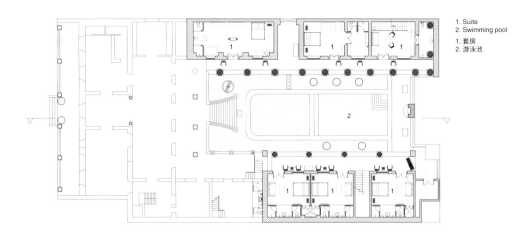

Ground floor plan (scale: 1/500) ／一层平面图（比例：1/500）

1. Suite
2. Swimming pool

1. 套房
2. 游泳池

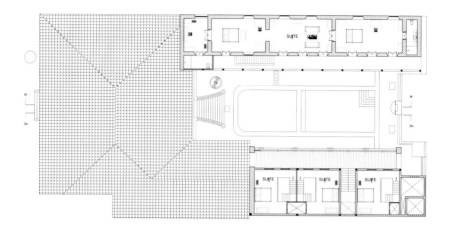

Top (mezzanin) level plan /顶层（阁楼）平面图

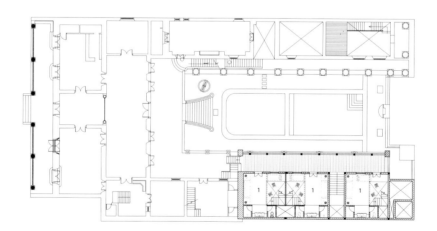

First floor plan(scale: 1/600) /二层平面图（比例：1/500）

加勒古堡酒店来自一座残破的住宅，它曾属于加勒本土一个名门商贾家族。其原始建筑可以追溯到 18 世纪，在 19 世纪和 20 世纪初为了彰显财富和公众形象而有了各种各样的加建。19 世纪末，随着科伦坡港口的发展，新的防波堤建成，业主家族将业务转移到了科伦坡，加勒的房子因此被空置废弃。该建筑最终被用作宝石加工厂，粗糙的原石在此被切割并制成闪亮的小玩意。1988 年，城堡区成为世界遗产。之后，人们对这里的兴趣日益加深，这里吸引了众多投资者购买并翻新此地的旧建筑，作为度假屋或酒店。这就是其业主回到这里的原因。

正如许多被废弃的古建筑一样，该建筑在经彻底打扫和初步拆除临时添加的墙与隔板后，便展现出一座宏伟住宅的气派。18 世纪的平面图展现了荷兰殖民时期的房屋样式。大阳台通向走廊，其两侧有两个房间，分别通向"扎尔"（荷兰语的大厅）。大厅朝向另一个可俯瞰后院的阳台，阳台外是一个美丽的双层柱廊区，顶部有低矮的主楼层房间，可能取代了原来的佣人区和厨房。据历史调查资料，这些新的加建部分是在 1920 年代为家族婚礼而做，其中包括入口大厅和大厅之间新奇非常的拱门。经过进一步清理后，卧室墙壁上人们发现了绘制于 1920 年代的壁画，这些壁画得以精心保存并被遮盖了起来。

对建筑的保护工作始于清晰记录现有肌理，然后根据城市发展局的保护规则进行规划。其过程还受到考古部门的监督。翻新工程想要增加一个不同于旧有时期风格的新馆，反映出不同时期的不同风格。整个建筑群变成一家小酒店，其中共建造出了 16 个房间，这使客人逗留在城堡区时能够感受到 18 世纪豪宅的风情。大多数室内设计都与当地的家具和织物配合。2007 年，该建筑由于在保护和再利用方面的卓越表现而被授予"联合国教科文组织亚太地区文化遗产保护"奖。

钱纳·达斯瓦特 / 文

Section (scale: 1/400) ／剖面图（比例：1/400）

Opposite: Louvered shutters provide privacy to the verandah of the 18 th century merchants' house.
对页：百叶窗为 18 世纪商人住宅的阳台提供了私密性。

This page, above: The interiors of the Hotel were conceived of by one of the owners, previously a filmmaker, who used his skills to reimagine life in this grand house. This page, below: Louvered shutters provide shade and privacy for the café. The chairs are replicas of an 18th century chair originally at the Wolfendhaal Church in Colombo. Opposite: When reconstructing the original ceiling of the central hall, timber planks on the edges were replaced with glass to allow skylight to penetrate into space.

本页，上：酒店的室内设计由一位业主（以前是电影制片人）构思，他用自己的专业重新想象这座大房子里的生活；本页，下：百叶窗为咖啡馆提供荫蔽和私密性。椅子是18世纪的复制品，在科伦坡的沃尔芬达尔教堂中被发现。对页：重建中央大厅的天花板时，边缘的木板被替换为玻璃，使光可以从天窗穿过进入室内。

Amila de Mel
No.5 @ Lunuganga (Osmund and Ena de Silva House)
Lunuganga, Dedduwa, Bentota 2013–2016

阿米拉·德·梅尔
卢努甘卡第五号住宅（奥斯蒙德与埃娜·德·席尔瓦住宅）
本托特，德都瓦，卢努甘卡 2013–2016

Credits and Data
Project title: No.5 @ Lunuganga (Osmund and Ena de Silva House)
Client: The Lunuganga Trust
Location: Lunuganga Estate, Dedduwa, Bentota, Sri Lanka
Original house completion: 1962
Reconstruction period: 2013–2016
Architect: Amila de Mel
Design Team: Amila De Mel, ADM Architects Pvt Ltd
Project Team: Nilan Cooray (conservationist); Deepal Wickramasinghe (structural engineer); Joe Fernando (quantity surveyor)
Project area: 627 m²
Project estimate: USD 165,000

No. 5 @ Lunuganga, also known as the Osmund and Ena de Silva House, was originally built as a private house in 1962 for Ena and Osmund de Silva. The client, in her own right a batik artist, had wanted a house that was rooted in the ancient architectural traditions of the country, but suited for a modern open lifestyle. In addition to the house was the location of a crafts cooperative and workshop initiated for women which was later continued by Ena de Silva in her home village of Matale, still running as a women's cooperative. The house was an inspiration to the workers and became the background to the initial sales as product of this initiative. Later in the 1980s, the house was rented by Geoffrey Bawa, the original architect, and used as a temporary office from which the drawings for the National Parliament of Sri Lankan designed by him with a specially recruited team of architects and designers were done; the house served as a secure environment for their work.

Soon after it was built, it was considered by many as a significant milestone in the development of modernism in the South-Asian or even larger Asian region. Published in many magazines of that time and eventually in books, it was one of the buildings central to the regionalist debate in modernism – used by critics such as Jim Richards to understand the significance of the modernist project outside its region of origin. For Sri Lankans in general, its architecture showed a way to modernity while connecting the possibility of using local building materials and skills in a time of austerity with government restrictions on imports of foreign goods, including building materials. In this way the building became an icon in the modern Sri Lankan design history.

The property has always been used as a residence and continues to be in its new location, albeit being moved to become part of a larger complex of buildings within the Lunuganga Estate of the late architect, Geoffrey Bawa. While the original site was a suburban one, here it is less so. However, the effectiveness of its architecture as a courtyard house is not compromised despite its new location. The major change over time was not really of the building itself, but of the original context that made the house unviable in its 50 years of existence. The house itself has a high wall which was originally shaded by a pergola on the street front. Within the site the building, there exists a single story structure with a garage, an entrance hall and a studio bedroom nearest to the street; a 2-story structure on the farther side with entertainment spaces on the ground floor and bedrooms on the upper floor separated by a large courtyard. The service and staff areas have spaces opening out onto several private courtyards that line the eastern side of the house.

Although the minister responsible placed a stay order on the demolition of the house at the outset, a public campaign held to save the house changed this situation and the protection status of the house in terms of it being a national criteria became unclear. It has no official status as a national monument, but, within the Lunuganga estate a management plan was developed to create internal protection. Conservation principles based on national and international standards are used to ensure its future survival; this includes creating access for the public to view the building so that it may continue to inspire the future generations of Sri Lankans.

Text by Channa Daswatte

---

p. 162: View of the central courtyard. The original frangipani tree did not survive the move; a tree very similar in shape and size is used in place of the original. p. 163: The satin wood columns, as seen in this photo, were made for the house in 1963. The original stone garden, setted up by Ena de Silva herself, is carefully rebuilt onto the new site. Opposite: The stone oil press that supports the roof of the entrance verandah are recycled from the original house to support the roof in its new location. Photos on pp. 162, 163, 170, 173 by Sebastian Posingis. Photos on pp. 164, 167, 169, 171, 172 by Nicholas Watt.

第 162 页：中央庭院。原来的缅栀花树未能在搬迁中保留下来，于是在新场地使用了形状、大小都非常相似的树来代替。第 163 页：1963 年为房屋建造的椴木柱。最早的石材花园由埃娜·德尔席瓦亲自建立，并被精心重建在新场地上。对页：支撑着入口走廊屋顶的石质压榨机被从原始房屋中回收，用于支撑新位置的屋顶。

卢努甘卡第五号住宅，又名奥斯蒙德与埃娜·德·席尔瓦住宅，最初于1962年作为埃娜和奥斯蒙德·德·席尔瓦的私人住宅而建。身为蜡染艺术家的客户埃娜·德·席尔瓦既想要一座植根于斯里兰卡古老建筑传统的住宅，也要求住宅能适合现代开放的生活方式。除了住宅功能外，此建筑还作为一个为妇女开设的手工业合作社和工作坊，后来，该组织由埃娜·德·席尔瓦在她的家乡马塔莱继续经营，今天仍作为妇女合作社运营着。这座住宅是工人的灵感来源，并成为了该项目销售的原型。在1980年代后期，该住宅被设计它的建筑师杰弗里·巴瓦租用，作为临时办公室。在这里他与特别招募的建筑师和设计师团队一起，为斯里兰卡议会大厦绘制了图纸。这里为他们的工作提供了安全的环境。

在这座建筑建成后不久，许多人便认为它是南亚乃至亚洲地区现代主义发展的重要里程碑。它出现在当时的许多杂志中，最后也以书籍的形式出版了。吉姆·理查兹等评论家将此建筑作为现代主义地域主义辩论的中心之一，用来理解现代主义项目在其起源地之外的意义。对于斯里兰卡人来说，在政府限制了包括建筑材料在内的外国商品进口时期，这一建筑提供了如何使用当地建筑材料和技能展现现代化建造方式的可能性。因此，该建筑成为了现代斯里兰卡设计史的象征。

尽管该建筑从原来的场地被搬迁至更大规模的建筑群，即已故建筑师杰弗里·巴瓦的卢努甘卡庄园中，它仍然同以前一样被作为住宅使用。相较于原场地，新场地不再那么郊区化，而建筑作为庭院住宅的建筑效果并未受到影响。时间流逝带来的主要变化实际并不在于建筑本身，而是使其在50年的历史中消失的原生环境。房屋本身有一堵高墙，被原场地街道前的藤架遮挡。建筑场地内有一个带车库的单层结构、一个门廊和靠近街道的单间卧室。远处还有一个双层结构，一楼有娱乐空间，二楼是卧室且由中庭隔开。服务和员工区通向排在房屋东侧的数个私人庭院。

虽然负责相关事宜的官员一开始下令拆除这座建筑，但为挽救它而举行的一场公众活动改变了这种情况，然而这座建筑在国家标准下的保护状况还不清楚。它没有"国家历史文物"的官方地位，但是在卢努甘卡庄园内，人们制定了一项管理计划以建立内部保护系统，使用基于国家和国际标准的保护原则以确保其未来的保存。这包括为公众提供参观建筑物的途径，以便它可以继续激励斯里兰卡的后代。

钱纳·达斯瓦特／文

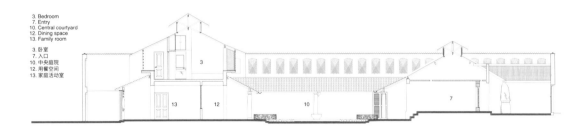

Section (scale: 1/250)／剖面图（比例：1/250）

1. Attic	6. Studio	11. Living room
2. Master bedroom	7. Entry	12. Dining space
3. Bedroom	8. Office	13. Family room
4. Toilet	9. Garage	14. Kitchen
5. Shrine room	10. Central courtyard	15. Pantry

1. 阁楼	6. 工作室	11. 客厅
2. 主卧	7. 入口	12. 餐厅
3. 卧室	8. 办公室	13. 家庭活动室
4. 洗手间	9. 车库	14. 厨房
5. 神龛室	10. 中央庭院	15. 食品储藏室

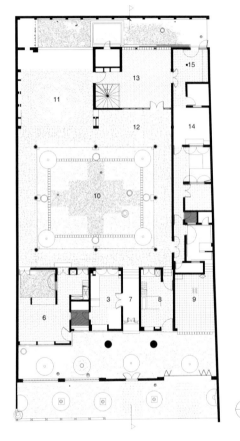

Ground floor plan (scale: 1/250) ／一层平面图（比例：1/250）

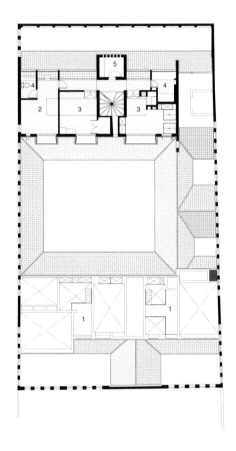

Upper floor plan ／上层平面图

p. 167: The courtyard of the artist studio is designed to allow natural light to stream into the room. Opposite: View of the original entrance door with its brass bells that were previously recycled from an old Hindu temple.

第 167 页：艺术家工作室的庭院使自然光洒进房间。 对页：原始入口门及铜钟，这些铜钟回收自旧印度教寺院。

This page, above: The original shutters and lattice work on the doors and windows are carefully conserved and reused as they were in their original locations. This page, below: The stone mosaic floor of the living room is arranged exactly as it was in the original house – a system of numbering and storage was used to ensure this. Opposite: The marriage of modern and tradition brings together the main structure made out of concrete and the courtyard verandah made from traditional materials.

本页，上：门窗上的原始百叶窗和格子细工都被精心保存和重新使用，如同位于原始位置上；本页，下：客厅石材拼花地板的布置与原始房屋完全相同，使用了编号和存储系统来确保这一点。对页：现代与传统的结合带来了混凝土主体结构和传统材料建成的庭院阳台。

Opposite: Each piece of timber parquet of this spiral staircase was carefully numbered, moved and reaffixed like an enormous 3D jigsaw puzzle. This page: The original lath and plaster ceiling did not survive during the move. However, with the help of the skilled craftsmen on site, a perfectly finished surface is created to accommodate the shape of the original column.

对页：这个螺旋楼梯的每一块实木地板都经过了仔细的编号、移动和重新固定，就像一个巨大的三维拼图游戏。本页：原始板条和石膏天花板在移动过程中没能保存下来。但是在现场熟练工匠的帮助下，创建出了完美适应原始柱状的表面。

Chapter 3:
An Environmental Ethic

第三章：
环境伦理

Islands are fragile domains. From their highly defined edges to their multiple centers, islands can be models of environmental responsibility as well as sites for hazardous behaviors. Sri Lanka is no different. Coveted for its abundance of unique sensitive environments but also linked to larger maritime fortunes and rebellions around the Indian Ocean, the island nation has seen itself as a semi-autonomous counterpoint to its larger, more industrialized neighbors. When considering how risk and resilience are forged within conceptions of architecture and urbanism on islands, there are few models that stand out in Sri Lanka. No sooner did the civil war conclude in 2009 than the building of skyscrapers, massive hotels and repetitive concrete structures throughout the country demand that access to fresh water sources, protected beaches and forests, and alternative fuel sources have become overlooked and translated as expensive commodities. The profusion of plastic in all forms has also laid waste to formerly pristine beaches, jungle paths and archaeological sites.

Each of the architects has adopted holistic approaches to the environment as an extension to responsibilities of architecture. Departing from standard material practices, the projects exemplify approaches to building with what may be found or discarded while using the ground itself as a component of a building's structuring. The projects disappear and in this way, allow one to reflect on impermanence as a technique for sustainable futures.

Sean Anderson

岛屿是脆弱的领域。它拥有定义明确的边缘和多重的中心，可以成为环境自运行的模型，但也是灾害频发的地方。斯里兰卡也是如此。这个岛国拥有令人梦寐以求的、丰富而独特的敏感环境，而且与印度洋广阔的海洋福泽和灾难联系在一起。因此，与更辽阔、工业化程度更高的邻国相比，他们认为自己是半自治的。然而在思考岛屿上的建筑和如何在城市主义概念中提升风险抵御力时，斯里兰卡只有很少的案例能脱颖而出。在2009年内战结束之前，全国各地的摩天大楼、大型酒店和重复的混凝土建筑都要求获得淡水资源、被保护的海滩和森林资源，替代燃料被忽视并转化为昂贵的商品。大量的各式塑料也污染了纯净的海滩、丛林小径和考古遗址。

在本章节收录的建筑中，每个建筑师都对环境采取整体应对方法，作为其建筑职责的扩展。与标准的材料实践不同，这些项目示范了如何利用可能被发现或丢弃的东西来建造，也将地面本身作为建筑物结构的一部分。这些项目消失了，但以这种方式可以让人们反思可持续未来技术的无常。

肖恩·安德森

Sunela Jayewardene
Jetwing Vil Uyana
Sigiriya 2002–2006

苏妮拉·贾瓦德
杰特维茵·维尔·乌亚那度假酒店
锡吉里耶 2002–2006

Positioning a new hotel, where all obvious niches were already filled, it was necessary to shift from the traditional approach to conceptualizing and designing a lifestyle experience. Situated on flatlands that stretch to the west of Sigiriya, a world heritage cultural site in central Sri Lanka, the design was considered an extension of an ancient irrigation plan done in the tradition of the Sinhalese kings as the basis of its approach. The dry and distressed agricultural land was fortuitously located at the end of a system of ancient reservoirs – a repository of cascading groundwater. The challenge was the effective conversion of agricultural land, formerly used for a traditional slash and burn method of cultivation, into a wetland. Incorporating the basic tenets of environmental architecture, the natural environment took precedence and the landscape is made the principal experience of the place. The buildings, inspired by vernacular architecture in the arid region, use timber frames with thatched roofs. The design details and placement of these structures react to the parameters of the new topography, acknowledging 4 distinct habitats: water, waters' edge, paddy field or marsh, and forest. Pivoting on water bodies, marshlands and woodlands, the contextual design of open pavilions, revives the ancient practice of pleasure gardens. The restored land – cultivated with indigenous flora – has been colonised by local fauna in a natural and ongoing process, with the fauna tripling and allowing onsite wildlife tourism. Jetwing Vil Uyana, through reclamation of agricultural lands for wildlife, is a global pioneer in "rewilding for tourism".

Text by Sunela Jayawardena
Edited by Channa Daswatte

Site plan (scale: 1/10,000) ／总平面图（比例：1/10,000）

pp. 176–177: Located at the end of a system of ancient reservoirs, the design intent was to extend the antique practice of irrigation into the site. This page: View of the main building designed around a water courtyard. Water is used as a form of passive cooling system for its spaces surrounding it. All photos on pp. 176–187 courtesy of Jetwing Hotels.

第 176-177 页：该设计位于古代水库系统的末端，旨在将古老的灌溉方法扩展到场地。本页：围绕水庭院设计的主楼。水被用作其周围空间的被动冷却系统。

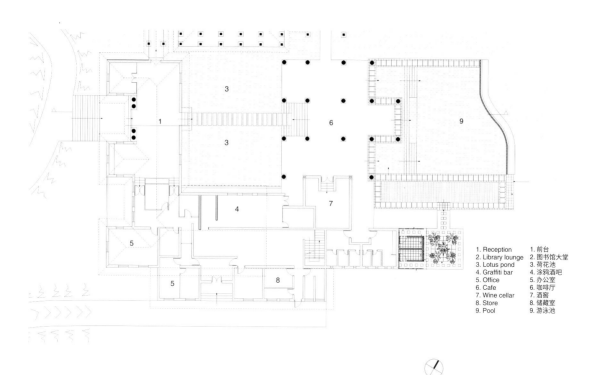

Plan of restaurant (scale: 1/500) ／餐厅平面图（比例：1/500）

Section of restaurant (scale: 1/500) ／餐厅剖面图（比例：1/500）

Opposite: The restaurant, in the style of a temple, opens up to views across to the UNESCO site of Sigiriya. A story-painting in vegetable dye can be seen in the photo background. pp. 182-183: View of the water chalets with coconut thatch-roofs, library, and restaurant across the lower reservoir and built wetland. p. 185, above: The forest cottage is built on high ground in a replanted forest. p. 185, below: Interior view of the forest cottage chalet. The bathroom is wrapped around a courtyard to provide natural light and ventilation into the spaces. p. 186, above: The interior of the paddy field chalet uses teak platforms and bamboo paneling. p. 186, below: The paddy field chalets are inspired from pavilions built by farmers to watch their fields.

对页：餐厅采用寺庙建筑风格，可欣赏到联合国教科文组织锡吉里耶遗址的景色。照片背景是一幅用植物染料绘制的故事画。第 182-183 页：横跨低层水库和人工湿地的椰子茅草屋、图书馆和餐厅。第 185 页，上：森林小屋建在重新种植的森林高地上；第 185 页，下：森林小屋内部。浴室环绕着庭院，以提供自然采光和通风。第 186 页，上：稻田小屋的内部使用柚木平台和竹镶板；第 186 页，下：稻田小屋的灵感来自于农民建造的、用来监看田地的凉亭。

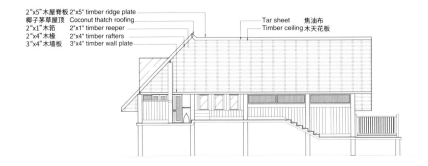

Section of paddy field chalet (scale: 1/200)／稻田木屋剖面图（比例：1/200）

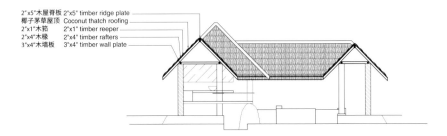

Section of the forest cottage (scale: 1/200)／森林小屋剖面图（比例：1/200）

当一个地区的酒店市场已明显饱和时，一家新酒店的定位就有必要从传统的方法转变为设计一种概念化的生活体验。杰特维茵·维尔·乌亚那度假酒店位于斯里兰卡中部、世界遗产文化遗址锡吉里耶西侧的平原上，其设计以僧伽罗皇家传承的古代灌溉计划为基础并进行了延伸。这片干旱贫瘠的农田碰巧位于一个古老水库系统的末端——一个倾泻的地下水库。设计的挑战是要将曾经用传统刀耕火种方式耕种的农用土地有效转变为湿地。结合环境友好型建筑要求自然环境优先的基本原则，景观塑造成为该场所的主要体验。这些建筑的设计灵感来自干旱地区的乡土建筑，采用了带有茅草屋顶的木构架。结构的设计细节和位置根据新场地的条件做出调整，确认了4个不同的栖息地：水、水域边缘、稻田沼泽和森林。开放式凉亭的设计以水体、沼泽地和林地为中心，将重现古代游园的实践。恢复后的土地种有本土植物，并在自然发展过程中成为了当地动物群的栖息地，而且当地生物总量增加了三倍，可以供游客观赏野生生物。杰特维茵·维尔·乌亚那度假酒店通过开垦农田为野生动植物提供生活区域，是全球"野生旅游"的先驱。

苏妮拉·贾瓦德 / 文
钱纳·达斯瓦特 / 编

Plan of paddy field chalet (scale: 1/250)
稻田木屋平面图（比例：1/250）

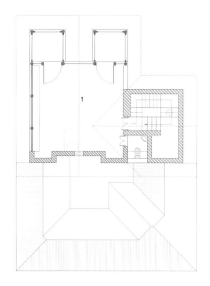

Upper floor plan of the forest cottage
森林小屋上层平面图

Credits and Data
Project title: Jetwing Vil Uyana
Client: Jetwing Hotels
Location: Sigiriya, Sri Lanka
Design: 2002
Completion: 2006
Architect: Sunela Jayewardene, Sunela Jayewardene Environmental Design Pvt Ltd
Project Team: C. Godaliyadde, S. Palugasweva (irrigation engineer); Stems Consultants Pvt Ltd, Jetwing Engineering (M&E engineer); Piyasoma Bentota (National Rush & Reed Federation, Sri Lanka); C. Panabokke (soil specialist); Dr. Sriyani Miththapala (biologist); L. Jayaratne (traditional muralist)
Project area: 10,700 m² (built area), 0.113 km² (site area)
Project estimate: USD 2,754,300 (500 million Sri Lankan Rupees)

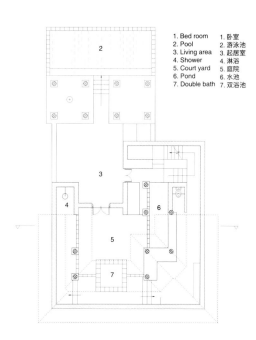

Ground floor plan of the forest cottage (scale: 1/250)
森林小屋一层平面图（比例：1/250）

L. A. R Kumarathunge, Tom Armstrong
The Mudhouse
Paramakanda, Anamaduwa, Puttalam 2005–

L. A. R 库马拉通加，汤姆·阿姆斯特朗
泥屋
普塔勒姆，阿讷默杜沃，帕拉马坎达 2005–

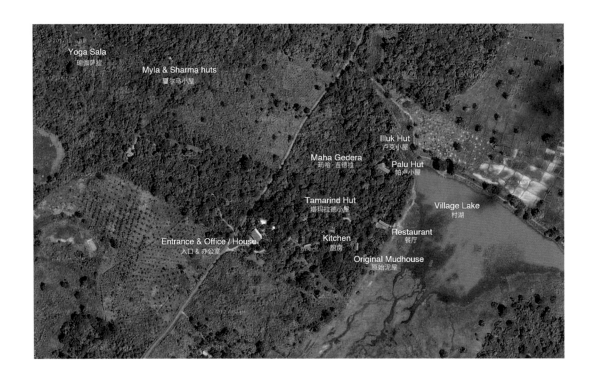

The Mudhouse is a forest lodge in rural Sri Lanka. It provides their guests with the chance to stay in an adventurous tropical accommodation that is built using a range of traditional techniques, albeit with some modern design elements. The Mudhouse is a nonprofit operation, with all its income being reinvested back into the project. Sustainability runs at the core of this venture, with environmental and social impact considered at every step. The Mudhouse has grown organically from a half-acre (2,023.5 m²) plot in 2005 to a 50 acres (202,350 m²) plot with 5 guest accommodation zones in 2019.

Text by L. A. R Kumarathunge and Tom Armstrong

Credits and Data
Project title: The Mudhouse
Client: self funded construction
Location: Paramakanda, Anamaduwa, Puttalam, Sri Lanka
Design: 2005
Completion: ongoing
Architect: L. A. R Kumarathunge
Design Team: L. A. R Kumarathunge, Tom Armstrong
Consultants: local builders and visiting professionals
Site area: 202,343 m²

p. 188: Site map. p. 189, from above and left: Yoga Sala. Sharma Hut, Maha Gedera, Original Mudhouse, Palu Hut. Opposite, above: The conical Yoga Sala sits majestically on a lake island. Opposite, below: The Nest, supported on stilts, has steps made from recycled railway sleepers. This page: Entrance to the Palu Hut. The hut is constructed using mud-ball walls, stone floor and uses furniture that were crafted in the Mudhouse workshop. Photos on pp. 188–195 by Damon Wilder.

第 188 页：基地图。第 189 页，从上到下，从左到右：瑜伽萨拉。夏尔马小屋，玛哈·吉德拉，原始泥屋，帕卢小屋。对页，上：圆锥形的瑜伽萨拉雄伟地坐落在湖岛上；对页，下：被高跷支撑着的巢屋，采用可回收的铁路枕木制成的楼梯。本页：帕卢小屋入口。小屋采用泥球墙和石地板建造，并使用在泥屋工坊中制作的家具。

泥屋是斯里兰卡乡村的一个森林旅馆。它为客人提供了一个可以住进热带住宅冒险的机会。旅馆设计上尽管具有一些现代设计元素，但主要采用一系列传统技术建造。泥屋是一个非营利性机构，其所有收入都被重新投资到该项目中。可持续发展是这项事业的核心，其建造的每个步骤都考虑到了环境和社会的影响。泥屋已从2005年的半英亩（2,023.5平方米）有机地发展到2019年的50英亩（202,350平方米），并且有5个住宿区。

L.A.R库马拉通加，汤姆·阿姆斯特朗/文

pp. 192–193: Interior of the Palu Hut, with a split-level coconut leaf thatch, mud floor and beds facing out into the forest. Opposite: View of a picnic hut overlooking one of the new water catchment channels that irrigates the on-site organic farm. This page, above: The Mudhouse organically grows a variety of rice grains in paddy fields jotted around the perimeter of the property. This page, below: No valuable trees or plants were cut during the hotel development. Paths between the huts are directed around the existing growth.

第 192-193 页：帕卢小屋内部。椰子叶葺成的不规则屋顶，泥为地面，面向森林的空间里安放有床铺。对页：一间野餐小屋，可俯瞰一个灌溉有机农场的新集水渠道。本页，上：泥屋在围绕场地周边的稻田中有机种植着各种水稻籽粒；本页，下：在酒店开发过程中，没有砍伐任何有价值的树木或植物。小屋之间路径的延伸遵循着现有环境。

Hirante Welandawe
Villa Santé
Kappalady, Talawila, Kalpitiya 2010-2014

希兰特·韦兰达维
桑特别墅
卡尔皮蒂耶，塔拉维拉，卡帕拉迪 2010-2014

Credits and Data
Project title: Villa Santé
Location: Beach Road, Kappalady, Talawila, Kalpitiya, Sri Lanka
Design: 2010
Completion: 2014
Architect: Hirante Welandawe
Design Team: Aparna Sen
Project Team: Nandana Abeyasuriya (structural engineer)
Project area: 557 m²
Project estimate: USD 500,000

pp. 196–197: Villa Santé is positioned in the midst of an existing coconut estate with two wings bordering the edges of the estate. This page: General view of the villa. Photos on pp. 196–203 by Ryan Wijayaratne.

第 196-197 页：桑特别墅位于现有椰子庄园的中间，其两翼靠近庄园的边缘。本页：别墅全景。

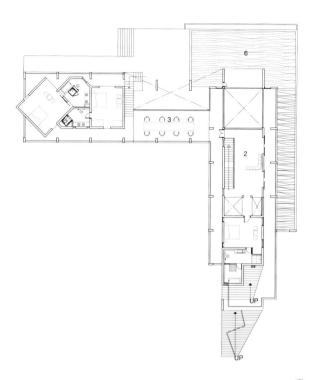

Upper floor plan ／上层平面图

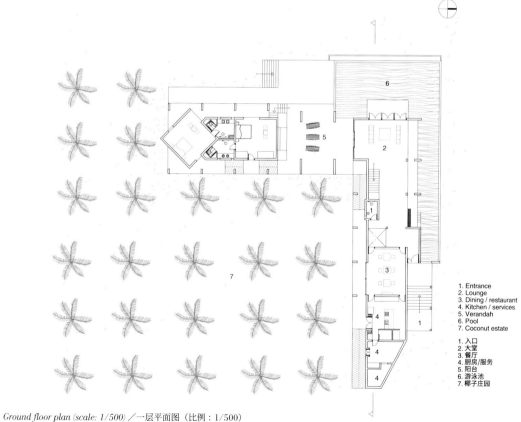

1. Entrance
2. Lounge
3. Dining / restaurant
4. Kitchen / services
5. Verandah
6. Pool
7. Coconut estate

1. 入口
2. 大堂
3. 餐厅
4. 厨房/服务
5. 阳台
6. 游泳池
7. 椰子庄园

Ground floor plan (scale: 1/500) ／一层平面图（比例：1/500）

Villa Santé is a boutique hotel located in a coconut estate in Kalpitiya, a western shorefront peninsular in Sri Lanka. Tourism in Kalpitiya is very different compared to the southwestern shores of the island – boasting marine life and wildlife that are unique to its shores. Thus, the act of pensive watching is as much a part of its activities as it is getting into the ocean. The steps of Villa Santé, an homage to Casa Malaparte, celebrates this act of pensive watching. The steps lead to a roof terrace hideaway with views of the expansive coast where one may sit and take in the surroundings.

The villa is positioned to fit in between the grid lines of existing coconut trees without disturbing the functioning coconut plantation. In fact, only one tree had to be displaced and replanted somewhere else during the construction process.

5 bedrooms, a public lounge and a dining area occupy the 2 concrete roof solid blocks placed in an "L" shape around a central court. All public areas are designed using passive ventilation and the materials used in construction are entirely locally sourced.

The leaf-shaped tiled roof is a design response to the tourism industry's enthusiasm for traditional Sri Lankan architecture; the tile roof, in particular, is considered an apolitical and exotic pre-existing architectural "other". It is, therefore, not only a roof covering the building, but also a decorative leaf straddling between two blocks. This separation of the roof and the built structure also allowed for the two elements to be treated differently; the built form as an earth bound and heavy object, and the roof as a lightweight structure perched on its timber composite columns. With this separation, the roof could also be experienced differently with its underbelly exposed, revealing its intricate network of timber members.

Text by Hirante Welandawe

Opposite, above: The roof shaped like a leaf floats above the solid concrete block massing that houses the private living spaces. Opposite, below: The steps of Villa Sante, drawing inspiration from Casa Malaparte in Italy, lead to a rooftop deck with a vast view overlooking the bay.

对页，上：屋顶像漂浮在坚实混凝土筑造的私人住宅上的树叶；对页，下：桑特别墅的台阶从意大利的玛拉巴特之家汲取灵感，通往屋顶露台，可俯瞰海湾。

This page, above: View from the first floor overlooking the verandahs provide an unobstructed visual connection to the bay. This page, below: The tiled roof, lightly supported by laminated timber columns, gently sits above two concrete blocks. The craftsmanship of the roof timberwork can be appreciated through this gap.

本页,上:从二楼阳台俯瞰,阳台为眺望海湾提供了无遮挡视野;本页,下:瓦片屋顶由层压木柱支撑,轻轻置于两个混凝土体块上方。通过这个间隙可以欣赏到屋顶木制品的精湛工艺。

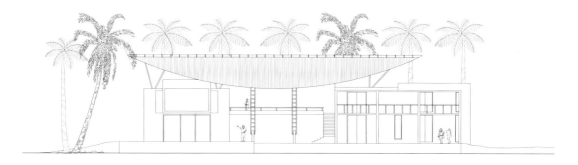

West elevation (scale: 1/400)／西立面图（比例：1/400）

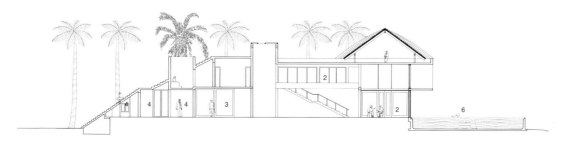

Section (scale: 1/400)／剖面图（比例：1/400）

桑特别墅是一家精品酒店，位于斯里兰卡西部海滨半岛卡尔皮蒂耶的一个椰子庄园中。与岛屿的西南海岸大不相同，卡尔皮蒂耶适宜发展旅游业，该岛以拥有独特的海洋生物和野生动植物而著称。因此，沉思、欣赏和亲近海洋一样，是这里活动的一部分。桑特别墅的台阶从玛拉巴特之家汲取灵感，进一步强化了游客沉思欣赏的氛围。这些台阶通向一个屋顶露台，在那里人们可以眺望广阔的海岸，坐下来欣赏周围的景色。

别墅被安置在现有椰子树之间，不会影响正在运营的椰子庄园。实际上，在建造过程中，仅有一棵树必须被置换并重新移植到其他地方。

5间卧室，1个公共休息室和1个用餐区占据了两座坚实的混凝土屋顶建筑。围绕中央庭院，它们被设计为一个"L"形。所有公共区域均采用被动通风设计，并且建筑中使用的材料完全来自本地。

树叶状瓦屋顶是对热门的斯里兰卡传统建筑旅游的一种设计回应。尤其是瓦屋顶，它被认为是一种非政治的、具有异国情调且不同于既有建筑的"外来者"。因此，它不仅是覆盖建筑的屋顶，也是跨在两个体量之间的装饰性叶子。屋顶与建筑结构的分隔也使这两个元素可以被区别对待。建筑整体结构相对笨重，而屋顶却是一个轻巧结构，栖息在复合木材柱之上。通过这种分隔，屋顶下方的可视化也可以提供不同的体验，展示复杂的木构系统。

希兰特·韦兰达维／文

Hirante Welandawe
Jaffna House
Thirunavely, Jaffna 2012-2014

希兰特，韦兰达维
贾夫纳之家
贾夫纳，蒂鲁内尔维利 2012-2014

Credits and Data
Project title: Jaffna House
Location: Thirunavely, Jaffna, Sri Lanka
Design: 2012
Completion: 2014
Architect: Hirante Welandawe
Design Team: Ahaladini Sridaran, Ayesha Kumari
Project Team: Nuwan Senaka (structural engineer)
Project area: 278.7 m²

The Public Private "Interspace"

Public and private are social constructs that conceptualize different domains of everyday life from the private domains of our homes to the public domains of our cities. The politics of the public and private define the architecture of this house. A citizen's house and the way they choose to dwell remains an important expression of their individual identity, which in turn contributes to a national character. The Jaffna House, for a keen observer and student of human behavior, expanded on that role and became a reflection not only of the individual occupant but also of the collective ideology and shared experiences of the community.

The House is located in the north of Sri Lanka, in an area that has been ravaged by a civil war for over 25 years. The recent peace has meant the picking up of the pieces by its people and the slow rebuilding of society.

When the brief came to Welandawe to design a residence in the heart of this war torn city, the vision for the design developed through conversations with the client and a gradual understanding of the outlook of the people of this region. The starting point was an insight towards the natural human tendency, under a consistent and systematic onslaught to reject the external and become introverted. The house became a reflection of this, envisioned as a sort of refuge from the devastations of the city. It became inward looking, with a blank inscrutable face that turned away from its surroundings; opening up instead, inside towards a green sanctuary. Another aftermath of devastating war is the very real fear of physical danger. Therefore, the security of the dwelling became an overriding concern, making the inward looking form a rational choice by enclosing and creating a retreat.

Greeted by blank walls with only a wooden fascia for relief, the residence seems sure-footed and stable resting on a solid base. The two floors of the house are arranged with the more public spaces on the ground floor and the private spaces, the bedrooms, resting lightly over them. Encased by high compound walls, the house reacts to the linearity of the site and elongated courtyards are created to frame the house, with the staircase forming an anchor guiding you through, leading to a hidden sanctuary, the garden in the back. While the house evades the external (the road and the city), it draws in nature (the rain, sun and sky) through this green space, deep into its recesses; as if turning its back on destruction and towards a more optimistic future.

The design also looked at the traditional houses of Jaffna that were in their essence, democratic spaces. Traditionally, an entrance verandah called "thinnai" is located at the front of the house, facing a small garden that in turn faces the street; this encourages gatherings and social interactions with people along the street. Some of these familiar spaces find an expression here, but Welandawe, as a concession to the changed time and circumstances, contests the role of these spaces and adapts them to suit their surroundings and the contemporary context – with the customary internal courtyards becoming extended spaces and pushed to the extremities of the house to create a buffer, and the typical street facing "thinnai" taken to the back of the house, now faces towards the private family garden. The design thus becomes an effort to find a valid architectural expression for the modern Jaffna lifestyle – keeping in mind their shared history and navigating the changing views of the people.

Text by Ahaladini Sridharan

---

pp. 204–205: The house opens to several private courtyards that provide natural ventilation, while also providing the family with the security and privacy that they need. Opposite: Interior view from the first floor verandah overlooking the double height living room. Photos on pp. 146–151 by Ryan Wijayaratne.

第 204-205 页：房屋通向数个私人庭院，在确保自然通风的同时也为家庭提供了他们所需的安全和私密性。对页：从二楼的阳台上可俯瞰双层高客厅的内部。

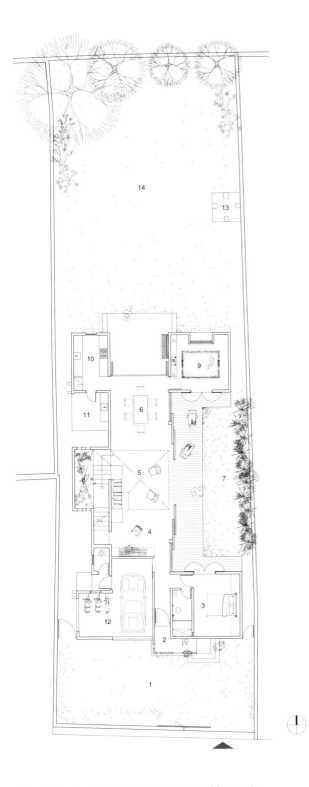
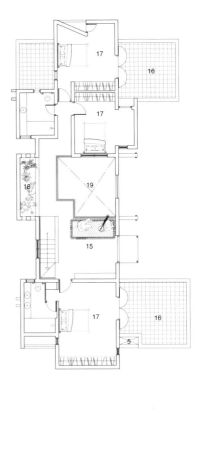

1. Yard
2. Entry
3. Bed room
4. Casual living
5. Formal living room
6. Dining
7. Family courtyard
8. Light well
9. Shrine room
10. Kitchen
11. Outdoor kitchen
12. Vehicle / scooter park
13. Well bathing / washing area
14. Garden
15. Study / music room
16. Terrace
17. Bed room
18. Light well
19. Void

1. 院子
2. 入口
3. 卧室
4. 休闲客厅
5. 正式客厅
6. 餐厅
7. 家庭庭院
8. 天井
9. 神龛室
10. 厨房
11. 室外厨房
12. 车辆／摩托车停车场
13. 浴池／清洗区
14. 花园
15. 学习／音乐室
16. 阳台
17. 卧室
18. 进光口
19. 空隙

Ground floor plan (scale: 1/250)／一层平面图（比例：1/250）　　*First floor plan*／二层平面图

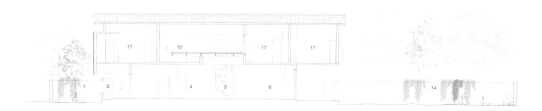

Long section (scale: 1/350) ／纵向剖面图（比例：1/350）

This page: General view of the house from the street. The front yard was intentionally left as a sandy courtyard, as it is still a part of the village homes in Jaffna.

本页：从街道看向房屋。前院被特意留为沙质庭院，因为它仍然是贾夫纳乡村住宅的一部分。

公私的"间隙"

就像社会由公与私构成一样，日常生活的不同领域也可以用这种公私概念来理解，比如说家是私人领域，城市是公共领域。公共和私人的政治关系定义了这所房子的建筑属性。公民住房及其居住方式仍然是个人身份认同的重要体现，反过来这又有助于形成民族特色。贾夫纳之家是人类行为的敏锐观察者和学习者，它深化着这一角色，不仅反映了居民个人，也成为了集体意识形态和社区共享经验的反映。

该建筑位于斯里兰卡北部。这里受到超过25年的内战的破坏，现今的和平意义重大，民众正在治愈伤痕，并慢慢尝试着重建家园。

当韦兰达维收到任务，要在这个饱经战火的城市中心设计住宅时，他通过与客户交谈，逐步了解这一地区居民的看法，随后提出了设计愿景。他的出发点源自对人性的洞察：当被不断抨击时，人就会开始拒绝外部刺激，甚至变得内向。该建筑反映了这一点，它被设想为远离城市惨淡情景的避难所。它是内向型的，对周遭环境保持沉默，但向内却开放了一个绿色庇护所。毁灭性战争的另一后果是造成对人身危险的真正恐惧。因此，住所的安全性成为首要关注的问题，而创造内向型的避难所也成为合理的选择，从而有了这座由墙围合起来的小隐之家。

迎接访客的是空白的墙壁，仅其木制饰物带来一些变化。该住宅立足于坚实的基础上，看上去稳固而踏实。房屋有两层，一楼有更多的公共空间，私人空间和卧室坐落其上。房屋被高大的复合墙所包围，与线性场地相呼应，并创建了细长的庭院来框定房屋。楼梯仿佛一个锚，引导人穿过，通往一个隐藏的避难所，即后花园。房屋本身与外部（道路和城市）保持着距离，通过绿色空间将自然元素（雨水、阳光和天空）引入屋内，仿佛抛弃了毁灭而走向更加光明的未来。

设计还着眼于贾夫纳传统建筑，它们本质上是民主化空间。传统上，入口的阳台被称为"提乃"，通常位于屋前并面对一个小花园，而小花园又面向街道。这鼓励了人们在街边聚会和社交互动。一些为当地人熟知的空间也在这里表达出来，但是出于对时间和环境变化的考虑，韦兰达维对这些空间的作用提出调整，使其适应周围和现代环境：例如，习以为常的内部庭院变成了扩展空间，并被移到房屋末端以创建缓冲区，而朝向街道的"提乃"被设置到房屋背面，面向私人花园。这样，设计便是在为现代贾夫纳生活方式寻找有效的建筑表达，既牢记共同历史，又引导人们不断变化的观点。

阿哈拉迪尼·斯里达兰 / 文

Opposite: View from the main garden positioned at the back of the house. A "thinnai", which is a verandah with a raised platform typically found in traditional houses, faces the private garden and is used as a casual seating or gathering space by the family and their friends.

对页：从位于房子后面的主花园看向房子。"提乃"是一个带有高架平台的阳台，通常用于传统房屋中，在这里面向私人花园，被用作家人和朋友休闲聚会的场所。

Channa Daswatte
Daswatte House
Madiwela, Sri Jayawardenepura Kotte 2002-2005

钱纳·达斯瓦特
达斯瓦特之家
科特,斯里·查亚沃登内布拉,曼迪维拉 2002–2005

The house was conceived as an open pavilion akin to the old ambalamas that dotted the Sri Lankan countryside as wayside resting places for travellers. The house is an open pavilion on the upper floor that rests on two solid blocks on the lower level. The pavilion, the main living space for its occupants, is surrounded by various kinds of shutters based on the level of privacy required for each area. Dense louvered shutters surround the library, bedroom and bathroom area, while light glass and steel shutters surround the main living space that opens to a view of the extensive gardens on both sides.

The solid blocks below house a staircase and kitchen on one side and a bedroom for guests on the other. The open space between is an entertainment space in the form of an open verandah that seamlessly becomes a covered park in the garden, similar to the verandahs of colonial bungalows where entertaining of its visitors took place. A vast dining table that seats twenty is in many ways the center of life in the house and its garden.

The entire house uses passive ventilation to keep it cool during most times of the year, except in high summer where the library and bedroom area may be air-conditioned by closing the glass and timber shutters. All other public and private spaces are always kept naturally ventilated.

Sustainability was an approach considered when deciding the materials used in this house and where possible, recycled building elements were used. The louvered shutters were recycled from Bentota Beach Hotel which was built by Geoffrey Bawa in 1967, and discarded during its refurbishment in 1995. Most of the doors, windows, and steel elements used were from other recycled materials. The rest of the new doors and windows were made using timber from trees that the architects' father cultivated, and later, more were planted in replacement of those that were used.

Text by Channa Daswatte

Credits and Data
Project title: Daswatte House
Client: Architect (self)
Location: Madiwela, Sri Jayawardenepura Kotte, Sri Lanka
Design: 2002
Completion: 2005
Architect: MICD Associates
Design Team: Channa Daswatte, Dilrukshi Paranamanage
Project Team: Joe Fernando (cost consultant); Mahesh Chandravansha (civil and structural engineer); Anzari Omar (contractor)
Project area: 60.4 m²
Project estimate: USD 120,000

pp. 212–213: View from the pool terrace facing the house with its open dining space on the ground floor, and its private living space on the upper floor. Photos on pp. 212–221 by Sebastian Posingis. Opposite, above: The guest cottage verandah flooring uses recycled handmade cement tiles, done in the tradition of Athangudi in South India. Opposite, below: The dining loggia opens seamlessly out onto the garden court. The stone flooring, recycled from a tea factory, enhances the outdoor quality of the space.

第 212-213 页：从泳池露台看向房屋。建筑底层有开放式就餐空间，上层是私人起居空间。对页，上：客用别墅的阳台地板采用可再生的手工水泥砖，这是印度南部阿唐加迪的传统；对页，下：餐厅凉廊无缝地通向衔接花园庭院。从茶厂回收的石材地板提高了空间的室外品质。

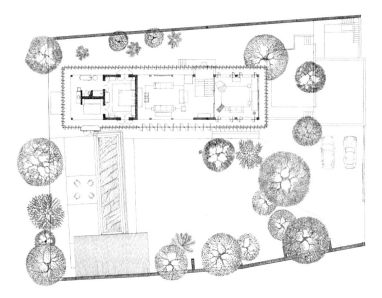

Upper floor plan ／上层平面图

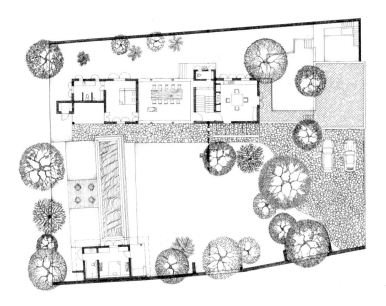

Ground floor plan (scale: 1/300) ／一层平面图（比例：1/300）

Opposite: The recycled cast iron bathtub that sits in the middle of the bathroom, the handmade batik ceiling, and the shutters are all recycled from Bentota Beach Hotel.

对页：位于浴室中间的再生铸铁浴缸、手工蜡染天花板和百叶窗均回收自本托特旅游度假村。

达斯瓦特之家被认为是一个开放式楼亭,类似于以前遍布斯里兰卡乡村、为旅人提供休憩场所的"Ambalama"(僧伽罗语"歇息之地"之意)。上层是开放式楼亭,位于下层两个坚实的封闭盒子之上。楼亭是居住者的主要生活空间,四周根据隐私级别安装有不同的百叶窗,以适应不同区域的需求。书房、卧室和浴室环绕着密集的百叶窗,而主起居室则环绕着轻质玻璃和钢制百叶窗,从屋内可以看到两侧开阔的花园。

下层的封闭盒子,一个设有楼梯和厨房,另一个设有供客人使用的卧室。两者之间的开放空间用作休闲的开放式阳台。也可视为花园中一个有顶的公园,类似于殖民地平房的阳台。宽敞的餐桌可容纳20人,在此可以招待访客,从各方面来说这里都是房屋及花园的生活中心。

在一年中的大部分时间里,整个房屋都通过自然通风来保持凉爽,除非是在盛夏,书房和卧室可以关闭玻璃和木材制成的百叶窗并使用空调。其他所有公共和私人空间始终保持自然通风。

可持续性是在确定房屋所使用材料时的一个考虑因素,并尽可能使用可回收的建筑废料。比如百叶窗是从杰弗里·巴瓦1967年设计的本托特旅游度假村中回收的,这些材料在酒店1995年进行翻新时被丢弃。大部分的门、窗和钢构件来自于其他回收材料。其余的新门窗均采用建筑师的父亲种植的树木制成,随后又种植了更多的新树来填补被砍伐的部分。

钱纳·达斯瓦特 / 文

Long section (scale: 1/200) / 纵向剖面图(比例:1/200)

pp. 218–219: *The living room on the upper floor uses pivoted glass shutters to provide natural light, ventilation, and unimpeded views of the garden. Opposite, above: To protect the books and antique piano, louvered shutters are used to modulate the light intensity entering the library, while the glass pivots placed between the cupboards allow the room to be shut for air-conditioning during humid weather. Opposite, below: Louvered shutters, recycled from Bentota Beach Hotel built by Geoffrey Bawa in 1969, provide privacy to the bathroom and dressing area while allowing natural ventilation.*

第218-219页:位于上层的起居室使用可旋转的玻璃百叶窗,以实现自然采光、通风和不受阻碍的花园景观。对页,上:为保护书籍和古董钢琴,百叶窗用于调节进入书房的光强度,同时置于橱柜之间的玻璃枢轴可以关闭,以便在潮湿天气下打开空调;对页,下:百叶门窗回收自杰弗里·巴瓦1969年设计的本托特旅游度假村,为浴室和更衣区提供了私密性,同时又保证了自然通风。

Amila de Mel
2B1 Container House
Mirihana Nagegoda 2004-

阿米拉·德·梅尔
2B1(集装箱住宅)
奴各果达,密里哈纳 2004-

2B1 began as an overgrown wild garden and gradually evolved into what it is now, a comfortable home that is part house part nature. Initially, there was no intention of making a house here in 1998. Two containers set 2.4 meters apart served as storage spaces. To make it habitable, a recycled warehouse roof at double height was set up over the two containers and this gradually transformed into an ephemeral jungle house with one container becoming a bedroom and the other divided into a kitchen and a bathroom. Later on, the containers were replaced by a brick-built successor that kept the original functions in precisely the same footprint under the warehouse roof. The bathroom has partial privacy, straddling the built-in space and the outdoors. The wooden deck overlooking the double height dining space, becomes a cozy and intimate space for entertaining while covering these rooms below. To keep the rain out, recycled glass shutters from a dismantled tea factory form an enclosure around the upper half perimeter of the warehouse structure. A double-size guest room was added later by the swimming pool at the bottom of the garden. Both the room and pool are hidden by the lovingly cared trees and foliage which now wraps around the entire structure. Despite the neighborhood now becoming part of the metropolitan Colombo area, waking up by the birds in the foliage, confronting the occasional chipmunks nibbling at the carpets, and enjoying fireflies at dinner are still possible in what is essentially a Garden retreat.

Text by Amila de Mel, edited by Channa Daswatte

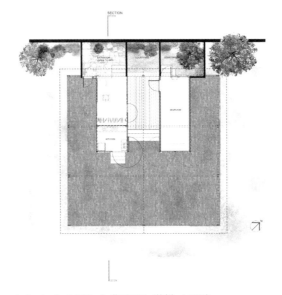

Phase 1 plan (scale: 1/200) ／一期平面图（比例：1/200）

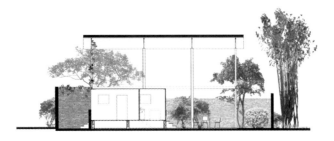

Phase 1 section (scale: 1/200) ／一期剖面图（比例：1/200）

pp. 222–223: The greens and climbing plants surrounds and envelopes the house, making its structure barely noticeable, blurring the boundaries between the inside and outside. This page, above: Two 6 meters containers, kept sheltered under a recycled warehouse roof, provided the containment for the original house. The container on the right was used for sleeping, and the left was used as the kitchen and toilet. The leftover open space function as a multi-use living space. This page, below: View of the container that was used as a kitchen and bathroom. Photos on this page by David Robson. Photos on pp. 222–233 by Sebastian Posingis unless otherwise noted.

第 222-223 页：观叶植物（译注：亦称观叶花卉，指一类叶形美好极具观赏价值的植物）和攀援植物围着房屋，令人几乎看不到建筑结构，从而模糊了内外之间的界限。本页，上：两个 6 米长的集装箱，一直被循环使用的仓库屋顶保护，仓库屋顶为原始房屋提供了安全壳。右侧的集装箱用于睡觉，左侧则是厨房和卫生间。剩余的开放空间用作多功能居住空间；本页，下：用作厨房和浴室的集装箱景观。

Opposite: General view of the living space from the exterior garden. This page: View of the living space.

对页：从外部花园看居住空间。本页：居住空间。

2B1最初是一个杂草丛生的野生花园，后来逐渐演变成现在的状态，即一个舒适的、半住宅半自然的家。1998年时，这里最开始没有打算建造房屋，而是用两个相距2.4米的集装箱作为仓储空间。为了便于居住，建筑师在两个集装箱上设置了一个双层高的再利用屋顶，这里逐渐转变成了一个临时丛林房屋，其中一个集装箱变成了卧室，另一个被设置为厨房和浴室。后来，砖砌房屋替代了这些集装箱仓储，但在仓库屋顶下的集装箱房屋被原封不动地保留了下来。浴室因横跨内置空间和室外，所只拥有部分私密空间。设计以木制甲板遮盖砖砌的箱体空间，从此处可一边俯瞰两层挑空的用餐空间，一边轻松惬意地招待客人。为了防止雨水进入，建筑师将从被拆迁茶厂回收的玻璃百叶窗再利用为仓库上部的围墙；之后，又在花园尽头的游泳池旁增建了一间两倍大的客房。房间和游泳池被环绕着整个建筑的树木遮蔽，这些树木都被精心照料着。尽管该社区现在已成为科伦坡地区的一部分，但本质上还是一个花园式的静修地。当使用者在树叶丛里、鸟鸣声中醒来，偶尔能看到花栗鼠啃食地毯，还可以在晚餐时看到萤火虫。

阿米拉·德·梅尔 / 文
钱纳·达斯瓦特 / 编

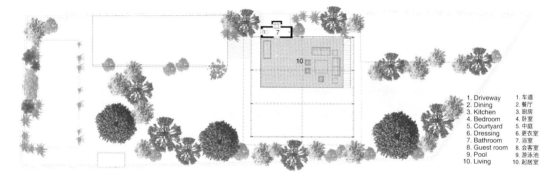

First floor plan／二层平面图

1. Driveway　1. 车道
2. Dining　2. 餐厅
3. Kitchen　3. 厨房
4. Bedroom　4. 卧室
5. Courtyard　5. 中庭
6. Dressing　6. 更衣室
7. Bathroom　7. 浴室
8. Guest room　8. 会客室
9. Pool　9. 游泳池
10. Living　10. 起居室

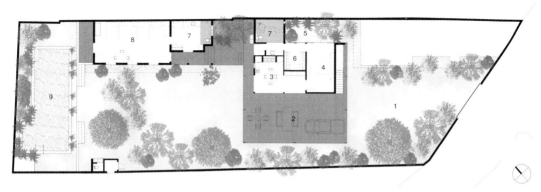

Ground floor plan (scale: 1/450)／一层平面图（比例：1/450）

Long section (scale: 1/400)／纵向剖面图（比例：1/400）

Credits and Data
Project title: 2B1 (Container House)
Client: Architect (self)
Location: Mirihana, Nugegoda, Sri Lanka
Design: 2004
Completion: Ongoing
Architect: Amila de Mel
Project area: 1000 m²

Southwest elevation (scale: 1/400)／西南立面图（比例：1/400）

pp. 228-229: The deck, supported by the two enclosed living spaces below, provides a form of retreat from the main living space, yet still keeping a close connection to its exterior gardens. p. 230: The brick and wood structure that houses the kitchen and bedrooms appear to be a terrace within a sheltered portion of the larger garden. The recycled tea factory window panels covering the upper perimeter of the house provide protection while maintaining a very subtle boundary between the inside and outside of the house. Photo courtesy of the architect. Opposite: View of the bedroom. Gentle sunlight enters the room through a courtyard located north of the house. This page: A breakfast table sits within an open kitchen, illuminated by dappled morning light.

第 228-229 页：由下方两个封闭式居住空间支撑的甲板提供了一种远离主要居住空间的形式，但仍与外部花园保持紧密联系。第 230 页：厨房和卧室的砖木结构似乎是较大花园中被遮挡部分的露台。回收的红茶厂窗板覆盖了房子的上部边缘，在提供保护的同时，也使房屋内外保持非常细微的边界。对页：卧室样貌。柔和的阳光穿过房屋北部的庭院进入房间。本页：斑驳晨光中开放式厨房里的早餐桌。

teaM Architrave
Office Building for Central Finance Co. Plc
268 Vauxhall Street, Colombo 2014-2017

阿奇特瑞弗建筑师事务所
中央金融有限公司办公楼
科伦坡，沃克斯豪尔路 268 号 2014–2017

Located next door to a 1970s office block for the same client by the late architect, Pani Tennekoon, the critical design challenge was to create both architectural and functional continuity in a contemporary statement – four decades later.

This project illustrates a vision for the City of Colombo. Unlike most construction happening in the city, this controlled and contextualized building complemented the street facade of the 70s building by adapting the same geometry to a 6-story, 3 meters deep steel and aluminium structure carrying luxuriant vegetation – a gridded vertical forest. The glass curtain wall is screened and completely shaded by this vertical forest. In another gesture, an internal courtyard in the older building is extended with a deeper court of the same width in the new building – creating a shared courtyard for the old and new buildings. At every level, dense planting continues across both courtyards.

Workspaces open out onto green aspects at every floor and at every position of the floor plan. Two decks are created to provide views towards the city (off the Board room and Auditorium) – unconventional practice for an office building. The 100-seat auditorium is located on the highest floor. Up-ending conventional practice, this is an outward-looking auditorium; it takes advantage of its sweeping city views by using a series of adjustable screens and blinds installed to make it possible to open out, whenever the opportunity arises.

The building employs extensive sustainability features: the reduction of air-conditioning loads via vertical planting and zoning of plans, the reduction of glare through vegetation, in-built screening and rainwater harvesting.

Text by Madhura Prematilleke.

Credits and Data
Project title: Office Building For Central Finance Co. Plc
Client: Central Finance Co. Plc
Location: 268 Vauxhall Street, Colombo
Design: 2014
Completion: 2017
Architect: teaM Architrave
Design Team: Madhura Prematilleke, Chamikara Moses, Lakmani Ratnayake, Harshana Jayakody, Gayathri Lindagedara, Lakdini Gamage
Project Team: Deepal Wickremsinghe, Arsath Sajeer (Deepal Wickremsinghe Associates/ structural engineers); Chandana Dalugoda, Nimal Perera,
Kosala Kambradeniya (MEP engineers); Lalith Ratnayake (VForm/ cost consultants); Galaxy Enterprises (main contractor); Alufab Pvt ltd, Univell, Built-Mech, United Safety, ETA Melco, Bubula Gardens, Colma Furniture, VS Information, Jinasena (sub contractors)
Project area: 1500 m²

pp. 234-235: General view of the office. The proposed design is a controlled and contextualized insertion into the city fabric to create continuity with the adjacent 1970s office block. Opposite, above: View of the shared open courtyard from the interior. Opposite, below: The shared courtyard provides comfort for its occupants working and also creates continuity between the old and new buildings. p. 239: View from the timber deck outside the auditorium. p. 240: View of the steel and aluminium gridded vertical forest which shades the glass curtain-wall and its interior. Photos on pp. 234–241 by Kesara Ratnavibhushana.

第 234-235 页：办公楼全景。设计方案是一种可控且情境化的、融入城市肌理的结构，以创造与邻近的 1970 年代办公大楼的连续性。对页，上：从内部可以看到开放式公共庭院。对页，下：公共庭院为居住者的工作提供舒适感，并在新旧建筑物之间形成连续性。第 239 页：会堂外的木材甲板。第 240 页：钢和铝的垂直网格，营造出森林般的外观，遮蔽了玻璃幕墙及其内部。

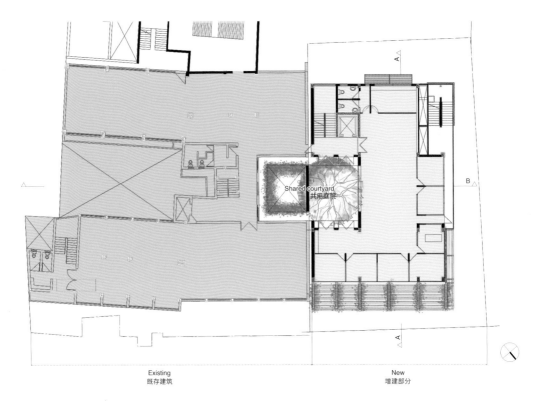

Typical floor plan (scale: 1/400) ／标准层平面图（比例：1/400）

这座建筑与邻近的一座1970年代的办公大楼同属一位业主，皆由已故建筑师帕尼尼·特纳库恩设计。此次设计的挑战在于如何在40年后的当代风格中创造建筑和功能的连续性。

该项目为科伦坡市描绘了一个新愿景。与这座城市中正在进行的大多数项目不同，建筑与周围环境密切相关，并利用相同的几何形状建造了一个6层高、3米深的钢和铝结构，其上承载着茂密的植被（形成垂直网格森林），从而对这座1970年代建筑所在的街道立面进行了补充。玻璃幕墙被垂直森林完全遮蔽。另一方面，旧建筑的内部庭院被扩展，一个具有相同宽度但更深的院落成为新旧建筑的共用庭院。在每一层，密集种植的植物景观都贯穿两个庭院。

工作区每个楼层和各个平面都向绿色空间打开。董事会议室和会堂外有两个平台，供人们欣赏城市景观，这对于办公楼来说是非常规的做法。最高层有一个100座的会堂，这个会堂以其开放性颠覆了传统惯例。它使用了一系列可调节的屏幕和百叶窗，确保空间可以随时打开，使人们可将城市风光一览无余。

该建筑具有广泛的可持续性特征：通过垂直种植和规划分区来减少空调负荷，通过植被减少眩光，同时内置过滤和雨水收集系统。

马杜拉·普莱马特雷克／文

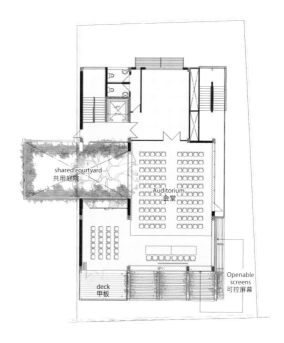

Auditrium floor plan (scale: 1/400) ／会堂层平面图（比例：1/400）

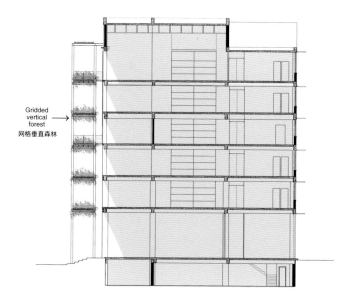

Section A (scale: 1/400) ／ A 剖面图（比例：1/400）

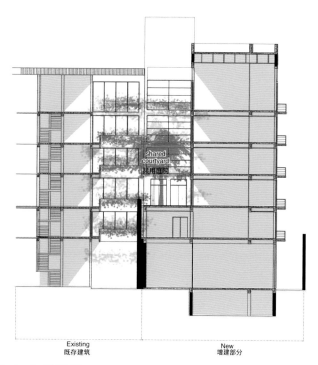

Section B ／ B 剖面图

Essay:
When is a Contemporary Sri Lankan Architecture?
Sean Anderson and Channa Daswatte

论文：
斯里兰卡建筑何时走向当代？
肖恩·安德森，钱纳·达斯瓦特

In 1947, on the eve of Sri Lankan independence when Geoffrey Bawa commenced practicing architecture, he began by reworking the landscape of a rubber estate into what would become his life's work: the gardens and home Lunuganga. This project recognizes the complex relationships found within the landscape and building throughout South and Southeast Asia as one of symbiosis, of shared communal values in a climate that determined daily rituals inasmuch as how one occupied space. This balance inspired the ways in which buildings might dictate movement, the passage of time, the materials used to structure space. Lunuganga became Bawa's crucible for experimenting with architectural ideas.

By 1998, and for most of the history of independent Sri Lanka, Bawa had been its most prominent and influential architect. A visit to Sri Lanka and Lunuganga by Charles, Prince of Wales, to mark the 50th anniversary of independence coincided with a terrorist bomb that destroyed the entrance of the 16th century Temple of the Tooth in Kandy. This reminds us that for the last 15 years of Bawa's life, his intertwined artistic circles and architectural practice had indeed flourished in the midst of a country at war with itself. While one may romanticize and sometimes even sanitize the elegance of buildings like the Kandalama Hotel (1992, *a+u* 11:06) or the House on the Red Cliffs (1997, *a+u* 11:06), it is difficult to reconcile the violence present in the country at the time with such beauty. When asked by a journalist during the war how he could go on making architecture when a country was in such a devastating state, the architect C. Anjalendran, responded that as an architect he was unable to physically participate in the war, but for all those people dying for the cause of the country, he could do his little bit by making respectable architecture and make the country a little bit more worth dying for.

By this account, late 20th and early 21st century architecture in Sri Lanka cannot dissociate itself from nearly thirty years of civil war that ceased in 2009. Unfortunately, the violence that at once sought division among the tight-knit societies of this island nation has since returned. Ten years following the cessation of hostilities, a new kind of intense aggression has transpired across the

1947年斯里兰卡独立前夕，杰弗里·巴瓦开始了建筑实践，他的第一个项目是重建一个橡胶庄园，卢努甘卡庄园。这个庄园后来成为了他的住宅和花园，也是他毕生的代表作。景观与建筑的复杂联系，在整个南亚和东南亚司空见惯，而该项目基于气候这个决定人们日常习惯和空间存在形式的要因，将其视作一种共生关系或价值共同体。这种平衡给建筑自身体量的延展、时间的流逝以及构成空间的材料带来了灵感。卢努甘卡成为了巴瓦试验建筑思想的熔炉。

至1998年，斯里兰卡独立后的大部分时间里，巴瓦一直是最杰出和最具影响力的建筑师。为庆祝斯里兰卡独立50周年，英国威尔士亲王查尔斯王子访问斯里兰卡和卢努甘卡，与此同时恐怖分子的炸弹摧毁了康提建于16世纪的佛牙寺入口。这提醒我们，在巴瓦晚年的15年中，他的艺术和建筑实践其实是在内战的情势下蓬勃发展的。尽管人们可能会将坎达拉玛酒店(1992年，a+u 11:06)或赤壁之家(1997年，a+u 11:06)之类的建筑环境浪漫化甚至纯净化，但这样的美与当时国家存在的暴力是很难调和的。建筑师切尔瓦迪莱·安贾伦德兰在内战时被一位记者问到，在国家处于如此灾难的时刻他怎么还能继续做建筑。他回答，那些为国捐躯的人们死得其所，而作为一名建筑师，虽然无法亲自参加战争，但他可以尽其所能地做出好的建筑，以此为祖国的建设出一份力。

因此，20世纪末至21世纪初的斯里兰卡建筑，与直到2009年才停止的近三十年内战不可分割。不幸的是，一度试图分裂该岛国的暴力活动已经开始回归。休战十年后，一股新的激烈攻击力量在世界各地蔓延，威胁着社会数年来的安全强化、经济自由和大型建筑。然而，多年的内战使得杰弗里·巴瓦和其他建筑师、艺术家和设计师都坚信战争终将结束。尽管总统府中新型高级安保设施被设计隐藏在景观之中，巴瓦自身却认为这类建筑并没有展示结束战争的希望，故而设计了一个最终效果为未完成状态的新建筑。对于一个根植于不间断的殖民镇压统治历史的国家，巴瓦以及21世纪初以来的建筑师的作品都蕴含着对未来的希望。

本书聚焦于斯里兰卡全岛自2003年以来的当代建筑。肖恩·安德森的导言讲述了那些定义独立初期建筑实践的先行者。先锋建筑师明奈特·德·席尔瓦、瓦伦丁·古那瑟卡拉和帕尼·特纳库恩通过使用进口材料和形式回应了超越国家概念的国际现代主义。他们的影响力立刻被背景各异、规模不同的项目吸收并例证。1960年代建立的教

world and upended years of improved security, economic liberalization and profligate building construction. Through the years of the civil war, however, Geoffrey Bawa and other architects, artists and designers anticipated an end to the fighting. Shown a design for a new high security complex embedded within a presidential residence buried in a landscape, Bawa reacted by saying that this kind of building had no hope for the war ever ending and went on to design a new structure for the purpose that remains unbuilt. For a country embedded within successive histories of colonial subjugation, Bawa, and those architects whose work has emerged since the turn of the 21st century, Sri Lankan architecture embodied hope for the future.

This book focuses on contemporary architecture in Sri Lanka with works throughout the island since 2003. Sean Anderson's introductory essay observes those antecedents that defined the early practices around the period of independence. Pioneering architects Minnette de Silva, Valentine Gunesekara, and Panini Tennakoon responded to an trans-national modernist ethos through their use of imported materials and forms. Their influences are at once absorbed by and exemplified among a number of varying scaled projects in distinct contexts. If pedagogical discourses established in the 1960s contributed to the image of a Sri Lankan architectural identity rooted in a history of discrete spatial experiences, they also provided an unyielding respect for locality, for the binding of structure to symbolic landscapes. This respect for locality through the use of available materials and skills continue to be a focus of many Sri Lankan architects today and while innovation in detail to meet contemporary sensibility is done by the designers themselves, engagement with highly skilled local craftsman have also made it a shared achievement.

More recently, the tenets of a globalized vernacular found among structures of expressive concrete, steel and glass are extended by an increased attention toward reexamining the limits and meanings of tradition. The architects featured here have sought to redress the distance between these conditions with buildings and landscapes that articulate connectedness across the nation but also a shared sense of responsibility in framing the island's view toward itself.

20 years after he stopped working, whilst celebrating the 100th anniversary of his birth in 2019, is this the future of hope in architecture that Geoffrey Bawa may have wanted to see?

学语境促成了植根于离散空间体验的斯里兰卡建筑特色，同时它也极其尊重对地域性、结构与象征性景观的结合。基于实际,使用当地材料和技术来尊重地域性，仍然是当今许多斯里兰卡建筑师关注的焦点。尽管为满足当代审美的细节创新是由设计师自己完成的,但也离不开当地技艺精湛的工匠的共同合作。

最近,随着对重新审视传统界限和意义的呼声日益增强,以表现力强的混凝土、钢和玻璃为结构的国际化乡土建筑的宗旨也被扩展了。本书介绍的建筑师们试图通过唤醒全岛共通的建筑与景观的和谐关系,来填补传统与国际化的鸿沟,同时他们也担任着塑造斯里兰卡自我形象的共同责任。

2019年,值杰弗里·巴瓦诞辰100周年,也是他停止工作的第20年。此时的建筑界,是他当年设想的未来吗?

Sean Anderson profile, refer to p. 31.
Channa Daswatte profile, refer to p. 246.

肖恩·安德森 简介，见第 31 页。
钱纳·达斯瓦特 简介，见第 246 页。

Architects Profile
建筑师简介

Channa Daswatte studied architecture at the University of Moratuwa and the Bartlett School of Architecture. He joined Geoffrey Bawa in 1991, before becoming a partner in 1997. In 1998, he formed MICD Associates with Murad Ismail, focusing on projects mainly in Sri Lanka, India, and other parts of the Indian Ocean rim. His project, Galle Fort Hotel, received an Award of Distinction in the UNESCO Asia-Pacific Heritage Awards of 2007. He is the Chair of Galle Heritage Foundation that coordinates efforts to conserve the UNESCO world heritage site. He has taught at both the City School of Architecture in Colombo and the University of Moratuwa, and has published several of his writings. He is a trustee of the Geoffrey Bawa and Lunuganga Trusts.

Portrait by Dhanushka Amarasekera.

钱纳·达斯瓦特曾在莫勒图沃大学和巴特莱特建筑学院学习建筑。他于1991年开始为杰弗里·巴瓦工作，1997年成为其合伙人。1998年，他与穆拉德·伊斯梅尔共同成立了MICD建筑师事务所，主要致力于斯里兰卡、印度及印度洋沿岸其他地区的项目。他的项目加勒古堡酒店2007年获得了"联合国教科文组织亚太遗产奖"。他是加勒遗产基金会的主席，该基金会负责协调保护联合国教科文组织世界遗产的工作。他曾在科伦坡城市建筑学院和莫勒图沃大学任教，并出版了诸多著作。他是杰弗里·巴瓦基金会和卢努甘卡信托基金会的受托人。

Amila de Mel began her training at art school in Boston, Massachusetts, prior to working with C. Anjalendran for 1.5 years. Later she joined and worked with Geoffrey Bawa for 5 years, principally on the Kandalama Hotel. In 1995, she furthered her studies at the University of East London, before returning to Sri Lanka to set up her own practice in 2000. Amila sits on the Board of Habitat for Humanity Sri Lanka, a nonprofit non-governmental organization, devoted to building "simple, decent, and affordable" housing. She is currently working with friends and colleagues to design low cost, sustainable, incremental housing models that are keeping with cultural background and heritage of the beneficiaries.

Portrait by Luxshmanan Nadaraja.

阿米拉·德·梅尔的专业学习始于美国马萨诸塞州波士顿的艺术学校，后为切尔瓦迪莱·安贾伦德兰工作了一年半。之后她加入了杰弗里·巴瓦的事务所，其工作的5年中主要负责坎达拉玛酒店项目。1995年，她前往东伦敦大学深造，于2000年返回斯里兰卡成立了自己的实践事务所。阿米拉是非营利性组织斯里兰卡仁人家园的理事，该机构致力于建设"简单、体面和可负担的"住房。她目前正在与朋友和同事合作，设计低成本、可持续并可复制的住房模型，以符合受益人的地域文化背景和遗产。

Hirante Welandawe graduated from the University of Moratuwa and obtained her Masters from the Alvar Aalto University in Helsinki. Her work follows a particular people centric way of approaching architecture focusing on socio cultural profiles and lifestyle, stemming from her early work which was concentrated exclusively on domestic and private spaces. She is a Visiting Design Tutor of the Colombo School of Architecture and has received several design awards and was nominated for the Aga Khan Award for Architecture in 2010. Welandawe's interest is in exploring the potential that Architecture offers to enhance the lived experience of the individual and the community.

Portrait courtesy of the architect.

希兰特·韦兰达维毕业于莫勒图沃大学，并在赫尔辛基的阿尔瓦·阿尔托大学获得硕士学位。她的工作遵循以人为本的准则，并注重社会文化特征和生活方式，这源于她早期主要专注于私人住宅的工作。她是科伦坡建筑学院的客座讲师，并获得了多项设计奖项，并于2010年获得阿卡汗建筑奖的提名。韦兰达维的兴趣在于探索"建筑"提供的潜力，以增强个人和社区的生活体验。

Madhura Prematilleke is an architect and urban designer based in Sri Lanka. He studied architecture in Moratuwa and Helsinki, worked and taught both in Sri Lanka and abroad. His practice, teaM Architrave (tA), Colombo has a commitment to urbanity and an ethos of crafted modernity, with a strong engagement in creating verdant tropical environments in restricted contexts, and a questioning approach to the paradoxes of consumption and sustainability. He has won 16 design awards and his work has been published widely: namely, the *Phaidon Atlas of Contemporary Architecture*, *Beyond Bawa*, *Architectural Review*, *Architectural Design*, and *a+u*. He is a regular speaker at international conferences, and juror for many international awards.

Portrait courtesy of the architect.

马杜拉·普莱马特雷克是斯里兰卡建筑师和城市设计师。他曾在莫勒图沃和赫尔辛基学习建筑，并在斯里兰卡国内外工作和任教。他的实践工作室阿奇特瑞弗建筑师事务所（tA）关注城市化和精心打造的现代精神，在有限的环境中大力创造葱郁的热带环境，并以提问的方法回应消费和可持续性的矛盾。他获得了16项设计大奖，其作品被广泛发表与出版，包括《the Phaidon Atlas of Contemporary Architecture》《Beyond Bawa》《Architectural Review》《Architectural Design》和《a + u》。他经常在国际会议上演讲并担任许多国际奖项的评委。

Trained in Colombo and Melbourne, **Milinda Pathiraja** (photo, left) received the RIBA President's Award for Outstanding PhD Thesis, CIOB Australasia Research Award, and the University of Melbourne Chancellor's Prize. Academically, he is attached to the Swiss Federal Institute of Technology (EPFL) in Lausanne and University of Moratuwa in Sri Lanka. In 2016, he was one of 88 architects to be profiled in the international section of 15th Architecture Biennale in Venice, Italy. His office, Robust Architecture Workshop (RAW) – led by **Ganga Rathnayake** (photo, right), Kolitha Perera and himself – won the Global LafargeHolcim Awards Silver prize in 2015, the regional Bronze prize for Asia Pacific in 2014, and the LafargeHolcim Building Better Recognition award in 2017.

Portrait courtesy of the architect.

米林达·帕蒂拉贾（照片，左）于科伦坡和墨尔本接受专业训练，获得了RIBA主席杰出博士论文奖，CIOB大洋洲研究奖和墨尔本大学总理奖。在学术上，他隶属于洛桑的瑞士联邦理工学院（EPFL）和斯里兰卡的莫勒图沃大学。他是2016年第15届意大利威尼斯双年展国际部分的88位最佳建筑师之一。他与甘加·拉特纳亚克（照片，右）和克利塔·佩雷拉共同领导罗伯斯特建筑工作室（RAW）赢得了拉法基霍尔西姆奖2015年全球银奖和2014年亚太地区铜奖，以及2017年拉法基霍尔西姆建造认可奖。

Palinda Kannangara is a Sri Lankan architect with a background in both mathematics and architecture. He graduated with a Bachelor of Science in Physical Sciences in 1996 with a specialization in Mathematics. During that time, in 1994, he joined a study course conducted by the Sri Lanka Institute of Architects. As a student, he trained under Sri Lankan modernist architect, Anura Rantavibhushana, who worked with Geoffrey Bawa for 16 years. Upon receiving his Charter in 2004, Palinda established his independent practice in 2005. His firm, Palinda Kannangara Architects located in Rajagiriya, Sri Lanka, is an award winning office known for an experiential architecture that hinges on simplicity and connection with the natural environment.

Portrait courtesy of the architect.

帕林达·堪纳卡拉是斯里兰卡建筑师，有数学和建筑学背景。他于1996年在物理系获得理学士学位，并主修数学。在此期间，他还于1994年参加了斯里兰卡建筑师协会举办的学习课程。作为一名学生，他师从斯里兰卡现代主义建筑师阿努拉·拉特纳布于曼纳，该建筑师曾与杰弗里·巴瓦合作16年。堪纳卡拉于2004年获得从业执照，并于2005年独立建立了自己的事务所。他的帕林达·堪纳卡拉建筑师事务所位于斯里兰卡拉贾吉里耶，是一家屡获殊荣的工作室，以注重简单性以及与自然相联系的实验建筑而闻名。

Sunela Jayewardene is widely recognised as "Sri Lanka's leading environmental architect" (*Time magazine*, March 2007; *India Today*, 2008), her primary design impulse is a serious concern for the ecology of sites and sustainability of human habitats. Environmental innovation is a central feature of her projects, with many award winning hotels and homes in Sri Lanka and India. Inspired by the traditions and landscapes of Sri Lanka, she seeks to preserve its culture by reviving crafts and vernacular design through applications in contemporary buildings. She is the principal partner of Sunela Jayewardene Environmental Design (Pvt.) Ltd. which focuses on concept development for environmentally sustainable architecture and land management projects.

Portrait courtesy of the architect.

苏妮拉·贾瓦德被公认为是"斯里兰卡环境型建筑的引领者"(《时代》杂志,2007年3月;《今日印度》,2008年),她的主要设计驱动是对场地生态和人类栖息地可持续性的关注。环境创新是她项目的核心特征,在斯里兰卡和印度拥有许多屡获殊荣的酒店和住宅。受斯里兰卡传统和景观的启发,她力求通过在当代建筑中,应用复兴手工艺和乡土设计来保存文化。她是苏妮拉·贾瓦德环境设计私营有限责任公司的主要合伙人,该公司专注于可持续建筑和土地管理项目的概念开发。

L. A. R Kumarathunge (photo, left), better known simply as Kumar, grew up in Kegalle District before moving to Anamaduwa aged 12. After a spell working in hotels and the family business, he decided to venture out with his friend, **Tom Armstrong** (photo, right), and start something that combined their shared enthusiasm for hospitality, adventure and design. They selected land in the rural heartlands of the country where they had met and slowly began on a journey that continues to this day. Despite having no prior architectural experience or training, Kumar was clearly a natural and was able to harness and develop his talent and channel his creativity in the years that have followed.

Portrait by Damon Wilder.

L. A. R库马拉通加(照片,左),更广为人知的名字是"库马",他在凯格勒区长大,12岁搬至阿讷默杜沃。在酒店和家族企业工作了一段时间之后,他决定与朋友**汤姆·阿姆斯特朗**(照片,右)共同闯荡,做一些可以结合他们爱好、冒险精神和设计的事。他们在相互认识的斯里兰卡乡村地带选择了一块土地,并慢慢踏上旅程,他们的旅程仍在继续。尽管没有任何建筑经验或培训,库马显然是一个天才,能够在随后的几年中利用并发展自己的才能,发挥创造力。

Thisara Thanapathy, born 1964, obtained his MSc in Architecture in 1992 from University of Moratuwa, Sri Lanka. He started his own practice in 1997, after working for Design Group Five (Pvt.) Ltd. for a period of three years and another two years at ADV consultants. He won the Geoffrey Bawa Award in 2011 and 2017 for Excellence in Architecture. His architecture maintains in embodying spaces that bring in relaxation, nourishment and rejuvenation from the stresses of modern day culture. His designs are often derived from their context and lies in balance with society, culture and nature.

Portrait courtesy of the architect.

蒂萨拉·泰纳帕里出生于1964年,1992年在斯里兰卡莫勒图沃大学获得建筑学硕士学位。他在第五设计集团私营有限责任公司工作三年,并在ADV咨询公司工作两年后,于1997年开始了自己的实践。他于2011年和2017年赢得了杰弗里·巴瓦卓越建筑设计奖。他的建筑坚持呈现在现代文化压力下休闲、滋养和唤回生命力的空间。他的设计经常来自于既有环境,诞生于社会、文化和自然的平衡。

Dilum Adhikari obtained her Bachelor of Science degree with honours in the field of the built environment and Master's degree in Architecture from the University of Moratuwa, Sri Lanka. She started as a Junior Architect at the State Engineering Corporation, before joining the Urban Development Authority (UDA) in 2005. She received her Charter in 2008, and since 2012, she has been working as an Assistant Director and Architect at UDA. Working at the UDA, she was able to major renovation projects like the Dutch Hospital Shopping Precinct which are now being considered as the spectacular Landmarks in Colombo city. Currently, she is working on Colombo Port City Development Project which is the first reclaimed city development project in Sri Lanka.

Portrait courtesy of the architect.

迪卢姆·阿迪卡里拥有斯里兰卡莫勒图沃大学建筑环境专业荣誉理学学士学位和建筑硕士学位。她最初在国家工程公司担任初级建筑师，之后于2005年加入城市发展局（UDA）。她于2008年获得从业执照。自2012年以来，她一直在UDA担任助理总监兼建筑师。在UDA工作期间，她主持了大型更新项目荷兰医院旧址购物美食广场，那里现在被认为是科伦坡市壮观的地标。目前，她正在负责科伦坡港口城市发展项目，这是斯里兰卡第一个重新开垦的城市发展项目。

Philip Weeraratne is the principal architect of PWA Architects in Colombo, Sri Lanka. Focusing primarily on design, his office values the process as much as the product. They do not claim allegiance to any specific design style or school of thought, and each project represents a new beginning. To them, architecture is the play of space, manipulating the inside to flow to the outside, the interplay of light and shade with appropriate use of volume and proportion to the available scale. Light and volume play on the senses of the visual and contact to create sensations and emotions within the user. To them, architecture is not a profession; it is a journey, where each project is a step towards a final destination.

Portrait by Earl Carter.

菲利普·韦拉拉特尼是斯里兰卡科伦坡PWA建筑师事务所的首席建筑师。他的事务所主要专注于设计，也重视设计结果和过程。他们不主张拥护任何特定的设计风格或思想流派，每个项目都代表新的起点。对他们来说，建筑是空间的游戏，通过适当使用体积和比例（与可用比例）以及明暗之间的相互作用，创造内部向外部的流动。光线和体积会在视觉和接触感官上发挥作用，从而启发使用者产生感觉和情感。对他们来说，建筑不是职业，而是一段旅程，每个项目都是通往最终目的地的一步。

Pulasthi Wijekoon (photo, below left), based in Sydney, is the Design Director for DG5i in Colombo. His design approach is innovative, inclusive, and rational with an inclination for sustainability through passive design strategies. **Guruge Ruwani** (photo, top left), now based in Sydney, brings clear conceptual direction, strong focus and balance with her holistic approach to design. She holds a fascination for materialising fluid, conceptual ideas through a rational process into rich spatial experiences. **Thusara Waidyasekara** (photo, top right), based in Colombo, is a principal architect of TWA who worked on a wide spectrum of projects. His career reached greater heights when leading his team during the Hambanthota International Convention Center project.

Portrait courtesy of the architects.

普拉斯丁·维耶空（照片，左下）在悉尼生活工作，是科伦坡DG5i的设计总监。他的设计方法新颖、包容、合理，并倾向于通过被动设计策略实现可持续性。现居于悉尼的**古鲁格·鲁瓦尼**（照片，左上）以其整体设计方法带来了明确的概念方向，强烈的关注和平衡。她着迷于通过合理的过程将丰富的概念性想法具象化为丰富的空间经验。科伦坡的**苏萨拉·威迪亚塞卡拉**（照片，右上）是TWA的总建筑师，从事过各种项目。他在主导设计汉班索塔国际会议中心项目中，到达了建筑事业的更高峰。

From the notebook of Shuji Kondo:
Memories of Geoffrey Bawa
Shuji Kondo

近藤秀次 记：
追忆杰弗里・巴瓦
近藤秀次

Among Japanese architects, I was probably the one who had worked with Geoffrey Bawa for the longest time. Starting in May 1979, I spent about half of the roughly three-year period until the project was completed in March 1982 – in other words, one-and-a-half years – working under Bawa on the design and supervision of the Sri Lanka Parliament Complex (1982).

The photograph, above right on the opposite page, was taken in May 1979 after a meeting we had in Colombo, during which I met Bawa for the first time, and discussed plans for the Parliament Complex. Bawa was really at the peak of his powers when he was 60 years old. At the time, I had just turned 30. Bawa is at the far left. On the far right is Dr. Poolgasundram, the structural engineer who also worked on the facilities. The lady in the center is Mrs. V. Jacobsen, a Sri Lankan architect who worked as an assistant to Bawa. I had meetings mainly with Bawa and Mrs. Jacobsen. The photo was taken in the lobby of the Galle Face Hotel, which was built during the era when Ceylon (now Sri Lanka) was still being governed by the Dutch. This elegant hotel was where the Showa Emperor stayed when he passed through Colombo on the way to Europe during the era of the Crown Prince.

Lunuganga, Bawa's country estate is located in Bentota where he would retire to each weekend. He would invite me to go to Lunuganga with him. There, I would dine together with him and talk about design while gazing out at the beautiful nature: all things that the design team would look very much forward to. The design was for the Sri Lanka Parliament Complex which needed to proceed at an incredible pace. We established an office that was separate from Bawa's residence in Colombo. We worked in a room with no air conditioning, but it was an environment where we could feel the comfort of the natural breeze. (Opposite, below)

Floor plans using 6m grid modules
Based on sketches that Bawa had made, we, the designers from the former Mitsui Construction Co., Ltd. (now Sumitomo Mitsui Construction Co., Ltd.) would copy them accurately onto tracing paper. These would be checked by Bawa while we produced a consolidated set of basic design drawings (floor plan, elevation, cross-section) from the ground floor to the fourth floor (5 floors) over a roughly three-month period. I went to Sri Lanka for the first time in May 1979, and subsequently remained onsite starting in mid-June, making drawings continuously from the morning until late at night.

Bawa's sketches had an extraordinarily accurate scale to them. (p. 252, both sketches) They seemed to have been drawn by laying them over a 6 m grid module. Once, when we were having a meeting, he turned to me and said, "Kondo, a grid measuring 20 feet by 20 feet! This will be perfect." So we erected columns at 6 m intervals, with a floor height of 5 m, while the Mitsui Construction structural design staff calculated the cross-section of the columns, and set the size of the cross-section at 60 cm by 60 cm. In Japan, this would come across as rather thin, but it was more than sufficient for Sri Lanka, where there are no earthquakes. Bawa treated proportion as something that was almost sacred. He took great pains to ensure that the size

我可能是日本建筑师中,和杰弗里·巴瓦(以下简称为巴瓦先生)工作过最久的人。我在斯里兰卡新议会大厦(1982年)项目中负责设计和监理的工作。从1979年5月起到1982年3月竣工的三年之间,我切切实实有一半的时间,也就是一年半是在巴瓦先生身边度过的。

本页的照片摄于1979年5月,当时我在科伦坡初次见到巴瓦先生,主要就斯里兰卡新议会大厦项目进行讨论。随后,我拜托他道,"请让我拍张照吧"。巴瓦先生60岁左右,正处于最意气风发的时候,那时我才刚满30岁。照片左边是巴瓦先生,右边是结构和设备工程师普罗卡桑德兰姆博士,中间的女性是斯里兰卡建筑师,也是巴瓦的助理建筑师瓦桑莎·雅各布森。巴瓦先生和雅各布森女士也是我的主要对接对象。拍照的地方是锡兰(斯里兰卡旧称)受荷兰统治时期建造的加勒菲斯酒店大厅。这是一家很有格调、比较高雅的酒店,昭和天皇还是皇太子时就曾在经由科伦坡前往欧洲时住宿于此。

卢努甘卡庄园(巴瓦先生的别墅)在本托特,巴瓦先生每个周末都会回到这个庄园,他也会邀请我,"近藤,到卢努甘卡来吧!"对于我们设计团队的成员来说,一边在别墅里用餐,一边眺望自然,一边谈论设计,是我们最期待的。我们不得不以极快的速度推进新议会大厦的设计工作。我们在巴瓦先生科伦坡的住宅旁边设有一个工作室。工作室里没有空调,但那里风和日丽,心旷神怡,环境令人感到愉快、舒适(本页,下图)。

采用6米网格模块的平面图

我们这些来自三井建设(现为三井住友建设株式会社)的设计师在描图纸上准确地抄绘巴瓦先生绘制的草图,然后巴瓦先生检查。在大约3个月内,我们完成了一层至五层的

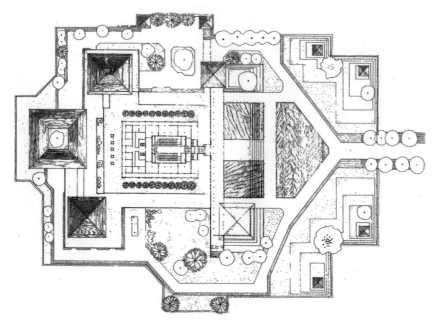

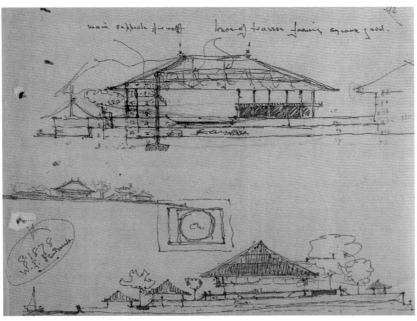

p. 251, from top: Photo taken at the lobby of Galle Face Hotel on May 1979. Photo taken at Lunuganga, Shuji Kondo is seen presenting his sketch to Bawa during a meeting. Photo showing the staff working on the design of Sri Lanka Parliament Complex, in the office located off Bawa's home in Colombo. All images courtesy of the author. This page, above: Bawa's sketch of the ground floor of the Sri Lanka Parliament Complex. This page, below: Bawa's sketch of the section of the Sri Lanka Parliament Complex.

第 251 页，从上开始：1979 年 5 月在科伦坡室内的加勒菲斯酒店。在卢努甘卡庄园里，近藤秀次正在向巴瓦展示素描。在科伦坡巴瓦自宅边的事务所里，员工们正在设计斯里兰卡新议会大厦。本页，上：巴瓦的素描，新议会大厦的一层。本页，下：巴瓦的手稿，新议会大厦的剖面。

平面、立面和剖面的扩初设计图。1979年5月我们初到斯里兰卡，6月中旬到达现场，之后就留在那里，早晚不停地绘制图纸。

巴瓦先生的草图比例非常准确，据他说，是将其放置在6米的网格模块上绘制的（对页，上下）。在平时开会时，他也对我说过："近藤，20英尺×20英尺的网格！就很好啦。"因此，我们在每6米处竖立一个支柱，使层高为5米，三井建设的结构设计师计算了支柱的横截面，并将支柱的截面设置为60厘米见方。这在日本感觉有点细，但是斯里兰卡没有地震，所以这个大小也足够了。巴瓦先生是一个非常看中比例的人，所以他非常讲究60厘米见方的截面，甚至包括最后的加工工艺层（结构本身的圆柱横截面因此变得更细）。于是，再次咨询结构设计师并重新计算后，我们终于圆满地确定了6米跨度和60厘米见方横截面积（包括饰面）。基于这些计算，我将所有的东西整合在一个6米的方格网格平面图上。巴瓦先生对此感到非常高兴，并对我说："近藤，你真了不起，几天的工夫就完成了这样的工作，如果是我事务所的员工，会花上十倍的时间"。从那时起，他就对我照顾有加，并每天与我讨论。

不过，设计进度仍然非常紧张，1979年11月时已经进入每个楼层的施工图设计阶段，而现场则开始以6米的网格打桩。也就是说，尽管施工设计图尚未完成，现场施工已经开始。这在日本是无法想象的。

新议会大厦的总体规划设计——自然与建筑共存

新议会大厦的规划基地位于科特行政中心的杜瓦岛（占地面积67,800平方米），距科伦坡市约10公里。当我们第一次拜访基地时，现场就像第258页照片中所示的，整块地都是湿地，感觉像是6到7公顷的浮岛。为了能够按照设计进行施工，一家荷兰公司对湿地的水进行了疏浚。最初，只有一条进场道路，为了应对突发状况，我们还建造了一条备用的道路。

我们的计划是无论如何都要实现巴瓦先生的设想。如果沿着总体配置图的轴线看，会有一种违和感，这是因为南侧的一栋建筑是错开的。在设计过程中，我们将绳子扎在现场并研究布局的可行性。巴瓦先生问道："那棵很大的山竹树怎么办？"我回答："因为它和建筑冲突，我正在考虑砍掉它"。巴瓦先生十分生气，并对我说"就算改变设计也绝对要留下这棵树！"因此，我们改变了设计，将南侧的建筑向南移动了6米（相当于一个网格），从而使这棵树位于建筑中庭的中心。

斯里兰卡的年平均气温为27°C至30°C。即使在这里，要长成这样的大树至少也需要30多年的时间。巴瓦先生教导我说，"以自然既有的姿态来设计建筑非常重要"。我和巴瓦先生一起查看了一圈现场生长的全部树木，他叮嘱我说，"不要砍掉这棵大树"。他还是一个善于开玩笑的人，"近藤，如果你砍了这棵树，你也会被砍哦"。

他是一位极为珍惜自然环境的建筑师。他常说："有大自然，才有建筑。因此，要充分利用自然的优势来做设计。"这个想法在他的许多建筑作品中都可以感受到。我认为最近日本年轻建筑师推崇巴瓦先生设计思想的最大原因是，他的设计坚持了"自然与建筑共存"的理念。

新议会大厦的天花板设计

我与巴瓦先生就新议会大厦天花板的设计曾争论不休。

巴瓦先生在桌上放了一块像手帕那样的白布，他用手捏住布的中间提起来，对我说道"近藤，我的概念是想做这样一个呈悬链状弯曲着的美丽的天花板"。他说，"我想使链网的中间看起来有被提起来、悬挂起来的感觉。"

在巴瓦先生这样说之后，我们研究了各种各样天花板的具体构想。如果在天花板上像网一样打孔的话，空调将无法运作，而且灰尘也会掉落，因此不能打孔。但从下面向上看时，如何使其看起来像是在网上开了洞？我的想法是在长宽各30厘米、高约3厘米的压印铝材料的表面进行压纹加工（第257页），把拐角切成锥形，做成便当盒的样子，之后再涂上黑色并将其贴在曲面胶合饰面板的基座上，这样或许可以使它看起来像是"网状图案"。

当从下方用红色灯光照明时，贴在黑色胶合板上的铝制便当盒会发出金色的光芒，天花板就会呈现出像链网那样的悬链曲线（第255页）。巴瓦先生对我的想法感到非常满意。

新议会大厦大屋顶的设计

斯里兰卡新议会大厦外立面的独特魅力在于拥有散布在古锡兰首都康提或日本古都京都和奈良众多古寺建筑的大屋顶，以及将木材用于外墙，使建筑与自然环境和谐地连接在一起的建筑手法。

在与巴瓦先生对接时，他提出的设计条件是：

of the cross-section was 60 cm, "including the finishes" (which meant that the cross-section of the structural column would become even thinner). After further discussion with the structural engineers and some recalculations, we were able to somehow maintain a 6 m span with a cross-section size of 60cm inclusive of finishes. Based on these calculations, I combined everything into a plan drawing on a 6 m grid. When I was done, Bawa was delighted. He told me, "Kondo, you're amazing. You took just a few days to finish this work. If it had been one of my own staff, they would have taken ten times as long." Ever since, I became his favorite, and we would have meetings every day.

The design schedule continued to be extremely tight, however. In November 1979, just as we were in the process of tackling the final design for each floor, the piling work began according to the 6 m grid. Even though the final design drawings had not been completed, the construction work went ahead at the same time. This would be totally inconceivable in Japan.

Overall design for the configuration of the buildings in the Sri Lanka Parliament Complex: symbiosis between nature and architecture

Duwa Island (site area 67,800 m²), in the administrative capital of Kotte, located about 10 km from downtown Colombo, was the planned site for the Sri Lanka Parliament Complex. When we first visited the planned site on Duwa Island, it looked like the landscape depicted in the photo on p. 258. The whole area was wetland, with around 6 or 7 hectares of islands floating within it. A Dutch company had done the work of dredging to remove the water from this wetland in order to shape the planned site. At the time, there was only one road leading up to the site, but we also created a line of movement from the back as a contingency plan.

As far as the plans were concerned, we made a concerted effort to attempt to realize what Bawa had sketched. One might feel a certain disjointedness in terms of how the axes come together in the diagrams for the overall configuration: this is because a single block to the south has been staggered to one side. During the design process, when we used ropes onsite to study the feasibility of the arrangement, Bawa asked me, "what we should do about that large mangosteen tree?" When I responded that I was thinking of cutting it down because it would get in the way of the building, he became extremely angry, and ordered me to change the design in order to keep the tree at all costs. Accordingly, I shifted the south block by 6 meters, the length of one grid, modifying the plans so that the large mangosteen tree appeared exactly in the middle of the central courtyard.

Even in Sri Lanka, where the average annual temperature is 27-30 °C, it takes more than 30 years for a tree to grow this big. Bawa taught me the importance of designing buildings while respecting and protecting nature as it is. He went around the site together with me to inspect the trees growing on it, checking to see that this or that large tree would not be felled. Bawa was quick to crack a joke as well. "Kondo, if you cut this tree down, I'll cut you off as well!"

Bawa was an architect who set great store by the natural environment. He was fond of saying that "architecture exists only because of nature. Design by making the best use of that nature, as far as possible." This philosophy is tangible in many of his architectural works. The main reason that Bawa's design philosophy has recently come to be appreciated by a young generation of Japanese architects is because it emphasizes the symbiosis between nature and architecture.

Design for the ceiling of the assembly hall of the Sri Lanka Parliament Complex

I grappled energetically with Bawa over the design of the ceiling for the assembly hall of the Sri Lanka Parliament Complex.

Bawa placed a square, handkerchief-like piece of white fabric on the desk, pinched the center and lifted it up while saying "Kondo, this is what I envision...I want to create a beautiful, catenary curve ceiling like this. As if you were pinching the center of a chain net and letting it hang down."

We were left to study various concrete ideas for the ceiling in response to what Bawa said. If

This page: View from the assembly hall of the Sri Lanka Parliament Complex, looking up to the ceiling. p. 257: The design of the aluminium bento boxes used on the ceiling.

本页：新议会大厦会议厅天花板。第 257 页：像便当盒的铝板天花板设计。

255

we opened holes in the ceiling like a net, the air conditioning would not work well. We couldn't make these holes, and dust would fall from them. What could we do to create the impression of holes in the net-like ceiling when one looked up at it from below?

I embossed the surface of square aluminium blocks measuring 30 cm across with a height of about 3 cm (p. 257), and cut off the corners to create a taper, creating something that resembled a bento box. Affixed onto a curved plywood base painted jet black, I reckoned that these blocks would appear to resemble a mesh pattern.

When these aluminium bento boxes affixed onto plywood panels painted black were illuminated from below by red light, they took on a golden glow, and a catenary curve-like ceiling that resembled a chain net seemed to appear on the actual ceiling (Opposite). Bawa was very pleased with my idea.

Design for the large roof of the Sri Lanka Parliament Complex

The unique charm of the exterior façade of the Sri Lanka Parliament Complex comes from the style of the large roofs of the numerous old temple buildings scattered throughout the ancient Ceylonese capital of Kandy, or the former Japanese capitals of Kyoto and Nara, as well as the methods by which wooden materials are used on the outer walls in order to connect the architecture to the natural environment, and create a certain harmony between them.

During a meeting with Bawa, he stipulated the following conditions for the design of the Parliament Complex.

1. The roof of the Parliament Complex would use roof tiles or copper plating.

2. As far as possible, natural teak would be used for wooden fixtures and surfaces.

3. We would like to use granite paneling for the foundations of the buildings. As the amount of granite that can be harvested in Sri Lanka is insufficient, however, a simulated stone finish will be used, made by turning granite into chips. Samples for this finish should be produced in Japan.

4. A steel frame structure will be deployed owing to the short construction period.

5. We would like to use Japan-made materials wherever possible, whether it is the cement, steel frames, steel rebars, or interior finishes.

As far as the roofing material of (1) was concerned, because the cross-sectional dimensions of the columns would necessarily become bigger as the roof load increases, after discussing the matter with Bawa, we decided to use copper roofs, subject to the following three issues being resolved.

A. We would create tile designs like those found on the roof tiles of the temples in the old capital of Kandy. Roof tiles in Kandy featured patterns with variations on deep or pale colors, forms and shapes, and the thickness of the tiles. We would create patterns for the copper plates, even for those used for the roof of the Parliament Complex.

B. We would bend the form of the big roof and design it so that it would feature a gently curving form.

C. The gaudy color of copper plates would be distasteful for a roof, so we would opt for a verdigris patina sprayed on upon completion of the building.

As such, regarding (A), the copper plates were folded to create a certain thickness, multiple patterns were created and affixed to the top of the flat copper plated roof, resulting in an effect close to that found on the roofs of the temples of the ancient city of Kandy. Regarding (B), the framework of the steel frames was folded and bent into three levels, creating a series of gentle curves on the roof. Regarding (C), the surfaces of the copper plates were chemically treated and oxidized before they were processed, so that the roof would take on a verdigris hue when the building was complete.

As it turned out, when the building was completed the big roof was a verdigris color, making this beautiful large-roofed piece of architecture blend into the surrounding natural environment and adjoining lake. Once again, I found myself thoroughly impressed by Bawa's sensibilities and exacting standards for design, glad that we had not ended up with a large roof with a gaudy copper color.

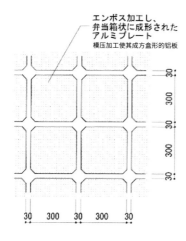

1. 新议会大厦的屋顶使用瓦或铜板。
2. 所用木材应尽可能为天然柚木板材。
3. 在建筑物底部(基坛部分)贴花岗岩。但是斯里兰卡生产的花岗岩数量不足,故而用由花岗岩制成的人造石材饰面。这个精加工样品希望在日本生产。
4. 结构采用钢结构,以减少工期。
5. 希望尽可能使用日本制造的材料,例如水泥、钢筋、钢框架,和内部装饰。

至于"1"的屋面材料,如果屋面载荷增加,则柱的横截面尺寸将不可避免地增加。因此,我与巴瓦先生讨论的结果是,在满足以下三点条件时,采用铜板屋面:

A. 设计类似于古城康提寺庙瓦屋顶的瓦片图案(康提的瓦屋顶使用不同的色调、形状和厚度的瓦片来给屋顶增加花纹)。在新议会大厦的屋顶上,使用铜板制作那样的图案。

B. 将大屋顶的形状弯曲,以形成柔和弯曲的效果。

C. 因为不喜欢铜板颜色的耀眼感觉,因此竣工时屋顶使用蓝绿色。

对于A项,为了增加铜板厚度,我们将其折叠并制作出多种纹样,然后将其粘贴在铜板平顶上,从而创造类似于古城康提寺庙屋顶的图案。对于B项,将钢制框架分三段弯曲,以创造屋顶平缓的曲线。关于C项,在处理铜板之前,对铜板的表面进行化学处理以使其氧化,从而在完成后可以实现蓝绿屋顶。

实际上,当建筑完工时,大屋顶已经变成了蓝绿色,融入了周围的自然环境和池塘景观,变成了一个美丽的大屋顶建筑。没有做成一个耀眼铜板色的屋顶真是太好了,我再次对巴瓦先生的敏锐和他对设计的考究感到由衷的敬佩。

对木材使用的考究及设计

虽然巴瓦先生要求我们在新议会大厦2至4层面向外部的走廊扶手、百叶窗及铜板屋顶的基底中使用天然柚木,但是工地负责人认为,如果所有木材都用柚木太困难了,需要从马来西亚和泰国进口。这不仅会使建造成本变得很高,同时获得柚木木材也会花费大量时间,可能会赶不上工期。于是,工地负责人给巴瓦先生展示了一块与柚木相似的低成本木材样品,并询问我们是否可以更换木材。巴瓦先生有些生气,并断然拒绝了这一提议。他说:"这种木材很容易腐烂,新议会大厦需要维持数十年。不是柚木绝对不行!"

巴瓦先生一直说"设计师是警察,施工者是强盗""承包商总试图诱骗设计师赚钱。"现场负责人拼命想说服他,以降低材料成本和建筑成本,却总是被巴瓦先生说服。

2017年,约在35年前建设新议会大厦的前三井建设(现三井住友建设株式会社)的工程师对斯里兰卡议会大厦的改建进行了实地调查。他们对我说,铜屋顶不会漏雨,外走廊的栏杆和百叶窗也没有腐烂。果然,巴瓦先生说的是对的,使用耐用的天然柚木真是太明智了。如果是其他木材肯定已经破烂不堪,朽坏腐烂了。巴瓦先生对建筑的

On Bawa's exacting standards for timber and its design

We were instructed to use natural teak for the handrails and louvers in the corridors facing the exterior on the first through third floors of the Parliament Complex. The construction manager, however, replied that it would be tough to use teak for everything. We would have to import it from Malaysia or Thailand, pushing up the construction cost. And time would be spent securing the necessary amount of teak, causing us to miss the construction deadline. We showed Bawa a sample of a wood that resembles teak but was a little cheaper, asking if we could possibly change the type of wood we would use. He got a bit upset: "the Parliament building is going to have to last for many decades. This wood rots easily. It absolutely has to be teak!" With that, he turned down the suggestion from the construction site.

Bawa always said that the designer was the policeman and the builders were the robbers. "The builders are always trying to cheat the designer and earn more money." Those in charge of the building on the construction site seek to keep the cost of materials down, even by just a little, in order to reduce the construction costs. They often fight back desperately, trying to persuade you one way or the other, but they always lost the argument to Bawa.

Two years ago (2017), a team of construction technicians from the former Mitsui Construction Co., Ltd. (now Sumitomo Mitsui Construction Co., Ltd.), the company that built the Parliament Complex around 35 years ago, conducted an on-site survey in preparation to renovate these buildings. They found that there were no rain leaks in the copper plated roofs, and that the teak used for the handrails and louvers along the external corridors had not rotted away. As one might expect, Bawa was right: thankfully, we had used hardy and durable natural teak. If it had been some other wood, it would undoubtedly have become dilapidated and rotted away. Bawa's exacting standards in architecture were not limited to the design: they were also applied to the choice of materials.

认真考究不仅体现在设计上,也体现在材料的选择上。

与巴瓦先生的回忆

巴瓦先生是一个重度吸烟者(真的是个无烟不欢的人)。即使开会的时候,他也总是在吸烟。我不喜欢烟味,所以对他说:"我不会坐巴瓦先生您开的车"。但他说:"来吧!近藤,别那么说,我们一起去吧",然后让我坐上了他的爱车。巴瓦先生喜欢和我一起开车去议会大厦的建筑现场。巴瓦先生有一辆丰田小型卡车,但他会开着经典的1930年代巧克力色劳斯莱斯,在时速约为15公里状态下翩翩缓行。我还记得当地的孩子看到巴瓦先生的车后,都会跑到近处来看一眼。

巴瓦先生在本托特的卢努甘卡庄园过周末时,经常什么都不做,只是凝望着周围的大自然。巴瓦先生的母亲非常美丽,他在卢努甘卡的卧室墙壁上挂着很多他母亲的照片。有一次,我问巴瓦先生为什么没有结婚,他笑了笑,但是没有回答。也许是因为没有遇到像他母亲那样美好的女性吧。

巴瓦先生一直都是单身,他只雇用年轻英俊的男子作为佣工。关于他的谣言从不间断。我自己这么说很冒昧,但那时我才三十多岁,长得也算可爱,所以建筑工地负责人经常对我开玩笑说,"比起你的设计能力,你的脸更被喜欢吧"。我真的每天都和巴瓦先生在一起,即使我回到日本,也多次收到他发给我的电报,邀请我快回去。

还有一个我和巴瓦先生都觉得有趣的故事。他告诉我,他在剑桥大学很用功地学习法律,然后成为了律师,主要负责离婚案件。但是,当他为顾客辩护离婚时,他们又会重新再在一起;反之,他建议双方再试一下,他们就会分开。从那时起,他就意识到他不适合当律师,于是在快40岁时加入了伦敦建筑学会,重新学习建筑,成为了建筑师。这

Opposite: Aerial photo of the site, Duwa Island, taken on May 1979. This page, from top: Perspective of the Sri Lanka Parliament Complex during basic design stage. Photo of the big roof made using copper plates, and finished with a verdigris color, taken after completion. Exterior view of the Sri Lanka Parliament Complex.

对页:鸟瞰杜瓦岛(摄于1979年5月)。本页,从上开始 在扩初设计过程中预期完成样子。青绿色的铜板大屋顶竣工时的样子。新议会大厦外观。

Memories of being with Bawa

Bawa was a chain smoker. He was always puffing away, even during meetings. I couldn't stand the smell of cigarettes. When I said that I wouldn't get into any car that Bawa was driving, he said, "come on, Kondo, don't say that, let's go together," and gave me a lift in his favorite car. Bawa liked to get in the car with me and go to the site of the Parliament Complex. He also owned a small Toyota truck, but usually he would be chugging along at about 15 km/h in a classic chocolate-colored Rolls Royce from the 1930s. I remember how the local kids would run towards the car to get a glimpse of his car whenever Bawa approached.

At Lunuganga, Bawa's country estate in Bentota, he would do nothing, gazing out at the vastness of nature surrounding his estate. Bawa's mother was an extraordinarily beautiful woman. The walls of his bedroom at Lunuganga were absolutely blanketed with photos of her. One day, I asked him why he never got married. He just smiled and said nothing. Perhaps it was because he never managed to meet a woman as wonderful as his own mother.

Bawa was a lifelong bachelor, and all of his servants were young, handsome men. There were all sorts of rumors about him, all the time. Perhaps this is somewhat presumptuous of me to say so myself, but I was still in my 30s at the time and good looking enough. The foreman at the construction site used to tease me: "I think Bawa likes your face, over and above your design talent." I really spent every single day with Bawa. Even when I returned to Japan, I would often get a telegram from him telling me to come back right away.

Another thing – a story that we both had a big laugh over. "I studied law seriously at Cambridge University, and became a barrister," Bawa told me. "I handled mostly divorce proceedings, but whenever I pleaded with my clients to get divorced, they would get back together again. Conversely, if I pleaded with them to reconcile, they would split up. It was then that I became thoroughly convinced that I was not cut out to be a barrister. So at close to 40, I enrolled at the Architectural Association in London and started over studying architecture, and became an architect." Perhaps the whole story was told in jest.

Books about Bawa often say something to this effect. "When he returned to Bentota from England, even when he sought out local designers to build his own house, they were mostly unable to produce what Bawa had in mind. His cousin told him, 'why don't you study architecture and design it yourself?' – and so he decided to become an architect." Bawa himself, however, would smile and tell me, "I devoted myself entirely to my legal studies and became a barrister, but I realized that I wasn't made for that profession, and that's how I became an architect."

More than an architect, Bawa was something of a philosopher. I was always astonished at how deeply he delved into things and understood them. Drawing on this knowledge, Bawa made his own designs for everything: the interior designs of various buildings, and even the design of the furniture inside them.

He was also a collector. His house in Colombo is now open to the public as a kind of "Bawa Museum." Most of the furniture and objects on display there were collected by the man himself. Bawa also had a penchant for old things. He would buy up large pots and urns standing next to a tree, or antique chairs. He would take with him things that caught his fancy not just from Sri Lanka, but also countries like India and Indonesia, and display them in his home. When he came to Japan to examine materials for tiles to be used in the construction of the Parliament Complex, he told me that he would love to take a Shigaraki ceramic tanuki (Japanese racoon) figure back with him to Sri Lanka, but I stopped him, saying that it would

Opposite: Photo of the Japanese gong donated to commemorate the completion of the building.
本页：竣工时本文作者近藤秀吉捐赠的日本吊钟。

也可能只是在开玩笑。

　　一部关于他的著作这样介绍道,他从英国回到本托特,请当地建筑师建造他自己的房子,但总是做不出巴瓦先生想象中的样子。他表姐就问他,要不要自己学习,然后设计。于是,他才以建筑师为目标。但是,他笑着对我说,"虽然我非常努力学习并成为了一名律师,但我意识到自己根本不适合当律师,所以我成为了一名建筑师。"

　　与其说巴瓦先生是建筑师,不如说他是一位哲学家。他对各种事物的深入研究和深刻理解总使我惊讶。基于这些知识,他还亲自设计各种建筑室内空间和家具。

　　他还是一个收藏家。他位于科伦坡的家现在作为巴瓦博物馆向公众开放,其展出的大部分家具和物品都是他的藏品。他对旧物没有任何抵抗力,像树旁的大罐子和瓮,以及古董椅子都是他买来的。不仅在斯里兰卡,他还从印度和印度尼西亚等地带回他看上的东西,摆放在家中。当他来日本检查用于建造新议会大厦的瓷砖时说,"我十分想将信乐烧的浣熊小雕像带回斯里兰卡"。我不得不阻止他,因为浣熊雕像很容易碎掉。巴瓦先生对此感到非常遗憾。

　　新议会大厦正门的走廊上悬挂着一口日本吊钟。有一天,他对我说,"近藤,你能送给我一口吊钟作为竣工礼物吗?"于是,我在日本富山县高冈市找到了"老子制作所",据说这是日本制作梵钟最好的地方。吊钟里用英文写上了"Dedicated to the friendship between Sri Lanka and Japan."(为斯里兰卡和日本的友谊)。我和巴瓦先生一起想了这个设计。我还记得我们两个画了各种各样的草图,商量着不要太日式,也不要像欧洲基督教的钟那样。可是,特别定制的设计花费的时间成本也高,并且无法在竣工时完成,所以变成了类似在日本寺庙中使用的钟的样子。

　　新议会大厦竣工时决定要拍照,我找到了新建筑社的摄影师。当我向他们介绍巴瓦先生的建筑后,他们不仅派

break easily. Bawa was terribly disappointed.

A Japanese bell, such as you might find in a temple, hangs in the corridor of the front entrance of the Parliament Building. One day, Bawa asked me if I might not make him a present of a Japanese gong to commemorate the completion of the building. So I had one made by the Oigo workshop in the city of Takaoka, Toyama Prefecture, widely considered to be the top manufacturer of temple bells in Japan. Written on the bell in English are the words "Dedicated to the friendship between Sri Lanka and Japan." Both Bawa and myself pondered over the design. I still remember how we produced various sketches together: we didn't want anything too "Japanese," and certainly not something that resembled the bells in Christian churches in Europe. As it turned out, a customized design would have been both time-consuming to produce and costly, and would not have been ready in time for the completion of the project. So the design ended up resembling something akin to the hanging bells in Japanese temples.

When the time came for photos of the completed Parliament Complex to be taken, I wanted to have a photographer from Shinkenchiku-sha to do it. When I showed them Bawa's architecture, they decided not only to send a photographer, but also to do a special issue (a+u 82:06) on the Parliament Complex as well as three hotel projects by Bawa. When I showed Bawa the published magazine, he said, "Japanese architectural photographers just take the most amazing photos. One isn't going to be enough, please send me ten more copies." "Kondo, thank you." I can still picture the smile on Bawa's face when he shook my hand. All I had wanted was to introduce the brilliance of Bawa's architecture to Japanese architects and students who aspired to enter the profession. I wanted the Japanese to realize that an architect of Bawa's stature existed in a small country like Sri Lanka.

The three years that I worked alongside Bawa were some of the best days of my professional life. Subsequently, I would go on to design projects in various other places like Singapore, Malaysia, and Taiwan, China. My three years of experience in Sri Lanka, however, were extremely significant. It was an era when materials were scarce, with many restrictions in terms of producing buildings, but I will always dearly cherish my encounter with Bawa and our many colorful conversations about architecture.

Translated by Darryl Jingwen Wee

了一位摄影师，而且还编辑成了"斯里兰卡新议会大厦·酒店3例"的特辑(a+u 82:06)。当杂志完成后，我把它展示给巴瓦先生。他说："日本建筑摄影师拍摄的照片很棒""一本不够，请再给我10本吧"。我现在仍然会回想起巴瓦先生和我握手并说"谢谢你，近藤"时的笑容。我当时的想法就是想向日本建筑师和有志于成为建筑师的学生们介绍巴瓦先生的杰出建筑。我想告诉日本的人们，在一个名叫斯里兰卡的小国里，有一位名叫巴瓦的出色建筑师。

与巴瓦先生共事的三年里，每一天都很美好。那之后，我虽然在新加坡、马来西亚和中国台湾等地区进行设计，但是在斯里兰卡的三年对我非常重要。在那个时代，供应物资短缺，对于建筑而言是一个不自由的时代，但能遇见巴瓦先生，谈论关于建筑的种种是我人生最大的财富。

Shuji Kondo was born 1949 in Hokkaido. In 1972, he graduated with a degree in Architecture from the School of Engineering at Hokkaido University. Upon his graduation that year, he joined the design department of Mitsui Construction Co., Ltd. (now Sumitomo Mitsui Construction Co., Ltd.). During the period May 1979 to March 1982, he was transferred to Geoffery Bawa's office in Sri Lanka to take charge of the Sri Lanka Parliament Complex. In 2000, he opened his own independent office, AA & Sun Associates. His recently completed projects include; covered corridor at East Promenade (Rinkai Fukutoshin, 2004), IKEA Funabashi (2005), VIA INN Nihombashi Ningyocho (2019).

近藤秀次 1949年出生于北海道。1972年毕业于北海道大学工学院建筑系。1972年加入三井建设（现三井住友建设株式会社）总公司设计部。1979年5月至1982年3月，为建设斯里兰卡新议会大厦与杰弗里·巴瓦的建筑事务所合作。2000成立自己的AA & Sun建筑师事务所。完成作品包括东海滨带屋顶走廊(临海副都心，2004年)，宜家船桥店（2005年），威亚酒店日本桥人形町（2019年）。

Bawa 100:
"The Gift": Artist-panel Discussion
Suhanya Raffel, Dominic Sansoni, Lee Mingwei, Shayari de Silva, Chandragupta Thenuwara, Dayanita Singh, Sean Anderson, Christopher Silva

巴瓦 100 周年：
"礼物"：艺术家圆桌论坛
苏安雅·华菲，多米尼克·桑索尼，李明维，沙耶里·德·席尔瓦，昌德拉古萨·特努瓦拉，达亚妮塔·辛格，肖恩·安德森，克里斯托弗·席尔瓦

Excerpts from an interview between Suhanya Raffel, Dominic Sansoni and Lee Mingwei.

Suhanya Raffel (SR): We framed the project at Lunuganga under the title "The Gift", because Lunuganga was a gift in every sense. To Geoffrey himself, it was an intellectual space, an aesthetic space, and an architectural space. A place where he problem-solved and thought about his practice. It was also a place of exchange where many artists and designers had worked with him, and so, the idea of "gift" felt apt in this year of Geoffrey's 100th birthday.

We will begin by talking about the work of Dominic Sansoni. In your work for the "The Gift", you're thinking of curating a selection of your photographs?

Dominic Sansoni: Yes. I'm going to try reviewing some of my very early pictures. But, there'll be some photographs which have never been taken yet, and so I'll be working, I fear, right up until the last minute. When you're given an assignment, there are two very different things that can happen. If it's somewhere new, you're excited as you've never been there before. You try to look in to discover and learn things, like meeting a new friend. And then, you also have old friends that you knew over the years and are familiar with. To me, what's interesting is to go back, see what you've done, and to observe and record some of that change.

(conversation continues...)

SR: Mingwei's practice is essentially about the relationships that he nourishes across places and seas. The (Bodhi tree) project, also a gift, brought him to Sri Lanka. It was a gift of a Bodhi Tree from Anuradhapura to the Queensland Art Gallery as an emblem of a new museum, to express that as a museum when you open, it is only just the beginning and with time you will grow into being what you will be.

In your project at Lunuganga, you brought an idea from one of your works, a set of tubular bells that were part of a "Trilogy of Sounds" made for Mount Stuart (17th century castle, Scotland). Could you share with us more about it?

Lee Mingwei (LMW): The process of bringing this project to Lunuganga from Scotland is actually quite simple. The tubes were produced in Sri Lanka with a specification and installed within about ten days? However, there were some challenges because the site was quite different in terms of weather, and the trees were different. In Scotland, we have four 400 year-old tall lime trees and they form a perfect square. But here, we only can find three trees. We had to change the design slightly while still creating a circular structure. That's the biggest challenge. Oh, and one other thing we have to fix is to convince the squirrels not to chew on the ropes.

(Mingwei describing his project...)

LMW: I was standing in the middle of the circle, and as I closed my eyes, I could feel the wind brushing against my face softly. Then, I hear the

苏安雅·华菲、多米尼克·桑索尼和李明维的访谈摘录

苏安雅：我们以"礼物"作为此次在卢努甘卡庄园的活动的主题，因为卢努甘卡从各个方面来说都是一个礼物。对杰弗里本人来说，这是一个知识空间、审美空间和建筑空间。这是一个他解决问题并思考个人实践的地方。这里也是那些与巴瓦一起工作过的艺术家和设计师交流的场所，因此，在杰弗里诞辰100周年之际，"礼物"是十分贴切的。

我们就从多米尼克·桑索尼的作品开始谈论吧。你准备选取一些你个人的摄影作品作为你的"礼物"吗？

多米尼克·桑索尼：是的，我将尝试回顾一些早期的摄影作品。不过，还有些地方从来没有拍过，所以恐怕我会一直工作到最后一刻。当艺术家接到任务后，可能会发生两种截然不同的情况。如果是新的东西，我们会感到前所未有的兴奋，会去尝试发现和学习新事物，就像结识新的朋友。还有一种是，与老朋友的再会。对我来说，追溯自己过去做了什么，并观察与记录自己的一些变化，偶尔也是乐事一桩。

(对话继续...)

苏安雅：本质上，明维的实践建立来自世界各地的人与人之间的联系。菩提树项目也是一个礼物，将他带到斯里兰卡。这棵菩提树是从阿努拉德普勒到昆士兰美术馆的礼物，是新博物馆的一个标志。它象征了博物馆的开馆只是一个开始，随着时间的流逝，它将会继续成长。

这里卢努甘卡的项目灵感是你从过去的一件作品中得来的，是在斯图亚特山(苏格兰17世纪城堡)制作的"声音三部曲"的一部分中的一组管状钟。可以与我们分享一下吗？

李明维：将这个项目从苏格兰带到卢努甘卡的过程实际上非常简单。这些管状钟是在斯里兰卡制作的，有详细的说明书，大约十天内就安装完成了。不过，这个项目还是有一些挑战，因为场地天气完全不同，树木也十分不同。在苏格兰，我们有4棵具有400年树龄的菩提树，它们形成了一个完美的广场。但是在这里，我们只能找到3棵。为了保持圆形结构，我们必须略微更改设计。那是最大的挑战。我们必须解决的另一件事是说服松鼠不要咀嚼绳索。

(李明维描述他的项目...)

李明维：我站在圆中间，闭上眼睛，能感觉到风轻拂着我的脸。然后，我听到钟开始唱歌，然后声音变得越来越强，越来越大。突然间，风开始散去，钟也停了下来。但是，声音仍然存在。你会听到它闪着微光，消失在森林中，声音跟随着风的脚步荡漾。

苏安雅：之所以将这一特别的作品带到卢努甘卡是因为要引出另一种要素，即声音，一种非常重要的感觉。它增强了人对花园的体验。

沙耶里·德·席尔瓦，昌德拉古萨·特努瓦拉，达亚妮塔·辛格之间的对话摘录

德·席尔瓦：首先，特努，你能简单告诉我们，你希望在肉桂山中探索什么吗？

265

bells start singing, and then, it gets stronger, and stronger, and stronger, and suddenly as the wind just started to go away, it stopped. But, the sound's still there. You could hear it shimmering and just disappearing into the forest. The sound follows the footstep of the wind.

SR: Another reason to bring this particular work to Lunuganga, was to draw on another element, sound, a very important sense. It amplifies your experience of the garden.

Excerpts from an interview between Shayari de Silva, Chandragupta Thenuwara and Dayanita Singh.

Shayari de Silva (SS): To start with, Thenu, could you briefly tell our audience what you're hoping to explore in the Cinnamon Hill site?

Chandragupta Thenuwara: At the cinnamon hill site, I saw bricks, many bricks. Unexpectedly, the patterns of the brick laying around an area reminded me of my work, the "Beautification" project. Then, I thought, I'm going to change that gravel area slightly. I'm going to continue the same brick patterns for a little bit, and äthen change it. And, I'm going to include the roots of the trees coming into my bricks as a three dimensional form. With these three different techniques, and the help of our ceramic department, I'm going to paint on these bricks, carve them, and at the same time, sculpt them.

(conversation continues...)

SS: Dayanita, I think the most important thing is the role of the photograph in your work and as a larger project. Could you elaborate on that, please?

Dayanita Singh (DS): The photograph has a very minuscule role in all of my work. I love saying it again, and again. The photograph is a raw material to me. It's just the starting of a work: it gives me the clues for where I might be able to take the work, or what form it can develop into. It's really just a raw material. In itself, it's not enough for me. I want much more out of photography. I feel I want to use every opportunity to push photography from under the rug – to dislodge the image and to make you think more about the image.

SS: You have been photographing and experiencing Geoffrey Bawa's work for many years. I was wondering if you could share with us about what his work means to you?

DS: I feel very connected to Geoffrey Bawa, and I don't regret that I didn't get to meet him in his lifetime, because I meet him continuously in some of his buildings. I think Bawa led me to architecture. When I finally got to Sri Lanka, waiting to meet with Channa, I started to recognize something in his aesthetic, especially when I went to Kandalama. And, I thought, "What is it? How do I know that reflection? How do I know that corner?". I haven't been to Sri Lanka, but I knew his work and I knew it very well. Then, I realized, through conversations with Channa, that he too, greatly admired the Padmanabhapuram palace

Opposite, from left: Photo of the interview between Lee Mingwei, Dominic Sansoni and Suhanya Raffel (from left to right). Photo of the interview between Chandragupta Thenuwara, Dayanita Singh and Shayari de Silva (from left to right). Photo of the conversation between Christopher Silva and Sean Anderson (left to right) about the Artists Installations in Lunuganga. Photos courtesy of Bawa100. This page: View of the Lunuganga where most of the Gift installations series will be held. Photo courtesy of the Lunuganga Trust.

对页,从左开始:李明维、多米尼克·桑索尼和苏安雅·华菲之间的访谈照片(从左到右);昌德拉古萨·特努瓦拉、达亚妮塔·辛格和沙耶里·德席尔瓦之间的访谈照片(从左到右);克里斯托弗·席尔瓦与肖恩·安德森(左至右)关于卢努甘卡艺术家装置对话的照片。本页:卢努甘卡庄园大部分"礼物"装置系列将在此展出。

outside Trivandrum, which is my favorite building in India. I understood we're all on a similar aesthetic wave length.

Excerpts from a conversation with Sean Anderson and Christopher Silva about the Artists Installations in Lunuganga.

Sean Anderson: When they asked me to come this evening, we were talking about what the role of temporary architecture might be in the context of a year-long exhibition that is extensively about art. We begin with Dominic's photographs, the work of making and imagining a garden. For Mingwei, the experience of that garden, through temporality and one's relationship to nature. Thenu, it was the material associated with politics. The ground that holds and supports, but also a source of great divide. Lastly, Dayanita, the notion of making one-zone time through image, process, or labour.

It is also very befitting to have Kuma design a temporary architecture – a pavilion – that will also be exhibited at the Lunuganga. In order to understand what the role of that temporal condition or architecture is, what I would do is a kind of palimpsest: a brief history of temporary architecture. To just step back, and think about why temporary architecture is important. Why is it significant to the history of architecture, and also more broadly, why we might be interested in showing temporary architecture in a garden like Lunuganga.

In the Barcelona Exhibition House designed by Mies van de Rohe, it was emblematic not only for what modern materials and architecture signified at that time, 1929, but also it signified what it could mean to live in this new, modern way. It was emblematic, then, of a system of values that was transforming. And when Geoffrey Bawa designed a temporary structure in Osaka in 1970, as an architect from Sri Lanka, he had to imagine what contemporary Sri Lanka represents.

I have the great privilege every year, as a curator, to organize the Young Architects Program, which started 20 years ago. This year, at MoMA PS1, it becomes a 3 months long architectural installation. It encourages each architect to think sustainably. For example, to think about light, about darkness, and even about accessibility to water. Ever since we started this program, architects have signaled ways in which we could think about temporariness and building.

Christopher Silva: Yes, and the Pavilion that Kuma has proposed to build in Lunuganga is going to be a 12 module structure. Kuma and his team were inspired by local kithul (made from Caryota urens, fish-tail palm tree) craftsmen and their craftsmanship, where they do amazing decorative and structural works with the use of kithul ran (the stem of kithul flower). It's a three dimensional weaving structure which you could see from both inside and outside.

Suhanya Raffel, Director, M+ Museum / Hong Kong; Trustee of the Geoffrey Bawa Trust and the Lunuganga Trust /Colombo. Dominic Sansoni, Photographer / Colombo. Lee Mingwei, Artist / New York and Paris. Shayari de Silva, Curator of Art and Archival Collections in the Lunuganga Trust / Colombo. Chandragupta Thenuwara, Artist / Colombo. Dayanita Singh, Photographer / New Delhi. Sean Anderson, Associate Curator in the Department of Architecture and Design at MoMA / New York. Christopher Silva, Architect, Associate Designer for Bawa100 / Colombo.

苏安雅·华菲，M+博物馆馆长 / 香港；杰佛里·巴瓦基金会和卢努甘卡信托基金会的受托人 / 科伦坡。**多米尼克·桑索尼**，摄影师 / 科伦坡。**李明维**，艺术家 / 纽约和巴黎。**沙耶里·德·席尔瓦**，卢努甘卡信托艺术与档案馆藏策展人 / 科伦坡。**昌德拉古萨·特努瓦拉**，艺术家 / 科伦坡。**达亚妮塔·辛格**，摄影师 / 新德里。**肖恩·安德森**，纽约现代艺术博物馆建筑与设计部门副馆长 / 纽约。**克里斯托弗·席尔瓦**，建筑师，巴瓦100的副总设计师 / 科伦坡。

特努瓦拉：在肉桂山遗址现场，我第一眼就看到了砖，很多的砖。出乎意料的是，围绕某个区域铺设的砖块图案使我想起了我的作品《Beautification》(《美》)。然后我想到，我要略微改造砾石区域。我将继续使用样式相同的砖，进行更改。我还将把树木的根部以三维的形式包括在砖块中。借助这三种不同的技术，在陶瓷部门的帮助下，我将在这些砖上绘画、雕刻并做造型。

(对话继续...)

德·席尔瓦：达亚妮塔，我认为你的作品中最重要的是将照片作为一个更大项目的核心。能详细说明一下吗？

辛格：照片在我所有的作品中都扮演着非常小的角色。我愿意不厌其烦地说明，照片对我来说是原材料。它是工作的开始，它为我提供了线索，可以带着我的作品去各个地方，或者可以发展成各种形式。但它实际上真的只是一种原材料。就我自己而言，这还不够。摄影之外我还需要更多。我想利用一切机会更深层地利用摄影：把图像赶出去并让人能够思考更深层的图景。

德·席尔瓦：您多年来一直在拍摄和体验杰弗里·巴瓦的作品，是否可以与我们分享他的作品对你的意义？

辛格：我觉得我与杰弗里·巴瓦之间有一种不解之缘。我并不后悔没有在他在世时与之会面，因为我与他在他的建筑中不断相遇。是巴瓦将我带进了建筑界。当我终于到达斯里兰卡、等待与钱纳见面时，我开始感受到巴瓦的审美观念。这在我到达坎达拉马之后更加明显。当时，我想，"这是什么？我怎么知道那个反射？我怎么知道那个角落？"我从未去过斯里兰卡，但是我知道他的作品，而且非常了解。通过与钱纳的交谈，我意识到他非常钦佩特里凡得郎之外的帕德马纳巴普拉姆宫殿，这也是我在印度最喜欢的建筑之一。我意识到我们的审美波长都差不多。

肖恩·安德森和克里斯托弗·席尔瓦就卢努甘卡庄园艺术家装置的访谈摘录

肖恩·安德森：这次他们来邀请我们时，我正在和我的团队讨论临时建筑在长达一年的艺术展览中，拥有怎样的角色和作用。我们从多米尼克的摄影作品开始，那是一个制作和想象花园的作品；而明维则是通过时间和与自然的关系来体验那个花园；特努使用了与政治有关的材料，这是固守与支持的基础，也是巨大分歧的根源；最后，达亚妮塔是通过图像、处理或人工来完成一个时区的概念。

隈研吾完全能胜任在卢努甘卡设计一座用于展览的临时展亭。为了理解临时性建筑的作用，我将要做的是类似于另行书写的事情，即书写临时建筑简史，也就是退一步来思考临时建筑为什么重要，为什么这对建筑历史具有重大意义，更广泛地说，为什么我们会对在卢努甘卡这样的花园中展示临时建筑感兴趣？

密斯·凡·德·罗设计的巴塞罗那展览馆不仅象征着1929年的现代材料和建筑，还指出了这种新的现代生活方式可能意味着什么。因此，它象征着蜕变的价值体系。当杰弗里·巴瓦作为一个斯里兰卡建筑师于1970年在大阪设计临时建筑时，他也不得不思考现代斯里兰卡代表了什么。

作为策展人，我有幸组织已有20年历史的青年建筑师计划。今年，在MoMA PS1上，它变成了长达3个月的建筑装置项目。它激励每个建筑师持续思考光、影、甚至亲水性。自从我们开始这个项目，建筑师已经提出了许多关于临时性和建筑的思考。

克里斯托弗·席尔瓦：是的，隈研吾设计的位于卢努甘卡的凉亭将由12个结构模块组成。隈研吾及其团队的灵感来自当地的基图尔(棕榈科鱼尾葵属董棕)手工编织工匠还有他们的手艺，他们使用董棕花茎进行令人惊异的装饰和结构工程。这是从正反两面都能看到的三维编织结构。

中日邦交正常化50周年纪念项目 日本国际交流基金会赞助项目

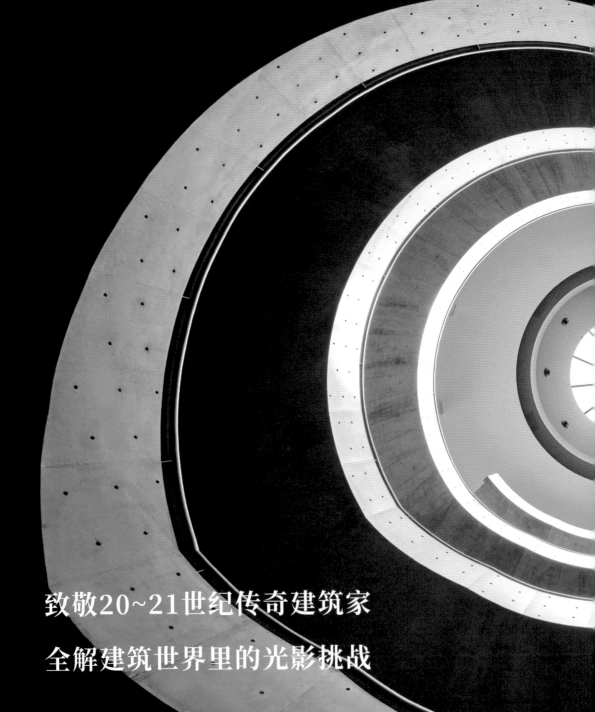

致敬20~21世纪传奇建筑家

全解建筑世界里的光影挑战

基本信息：开本 16开/尺寸 215mm×280mm/语言 内文全中文，索引中日英三语/页码 3496页（
筑联盟/主编 马卫东/执行编译 安藤忠雄全集编辑部/出版发行 中国建筑工业出版社/主要作(译

安藤忠雄全集
TADAO ANDO COMPLETE WORKS

380 +　建筑作品
5 +　　家具艺术品
1500 +　摄影作品
500 +　 手绘作品
100 + 论文及安藤故事
10 +　 展览及建筑小品

(准)结构 六卷/印刷 四色全彩/装帧 精装/发行范围 全球中文/监修 安藤忠雄建筑研究所/特别支持 日本新建筑社/书籍策划 文筑国际 IAM 国际建尼斯·弗兰姆普顿 铃木博之 彼得·艾森曼 松叶一清 田中川武 三宅理一 朱涛 马卫东等/本书相关消息敬请关注官方微信公众号"安藤忠雄之家"

© 和美术馆

Spotlight:
Liangzhu Center of Arts
Tadao Ando Architect & Associates

特别收录：
良渚文化艺术中心
安藤忠雄建筑研究所

The site is located in Liangzhu Culture Village on the outskirts of Hangzhou, less than 200km from Shanghai. It is a comprehensive facility designed to connect local residents with art, culture and education. The base has a long-standing reputation as the discovery place of the Archaeological Ruins of Liangzhu City, which were found in the lower reaches of the Yangtze River at the end of the Neolithic Age 4,000 to 5,000 years ago. Liangzhu culture formed a huge city with walls. Plows were used in agriculture to cultivate the land, and crops such as rice were grown at the time. Not only stone tools, but also pottery, jade, bronze and other cultural relics were found on the site. The Archaeological Ruins of Liangzhu City plays an important role in the exploration of the ancestors of Chinese civilization and culture.

The key focus of the building is to restore the rapidly lost regional community in Hangzhou, a city that has seen extraordinary economic growth. I tried to construct a "place for dialogue" that transcended the scope of simple cultural facilities, and form a space under the "big roof" that symbolizes families and regional communities. The roof, though buried in the forest, still serves as a landmark to express the character of the building.

In addition, considering that the West Lake and the surrounding canals tie the heart of Hangzhou people, I introduced the river near the site into the building through the water court, and planted cherry blossom forests between the water feature and the river. In spring, the foreground of the big roof by the water will be covered with a vast expanse of cherry blossoms.

The building is divided into three volumes according to functions: the gallery building, the cultural facilities building and the educational facilities building, which are arranged under the big roof. In addition, by offsetting the cultural facilities building in the center, the relationship between the blank space under the eaves between the three volumes and the quiet water scenery in the foreground of the building becomes full of diversity. The space under the big roof has become an external space that triggers people's activities such as dramas, concerts and fashion shows. My goal is to maximize the use of the site, to achieve a "dialogue with riparian landscape" and to create a good environment for human to contact with culture.

Text by Tadao Ando

Credits and Data
Project title: Liangzhu Center of Arts
Location: Hangzhou,China
Design: 2010.7-2013.4
Completion: 2013.5-2015.7
Structure: Reinforced Concrete, Steel
Fuction: Multifunction
Site area: 23,558.00 m^2
Building area: 4,946.00 m^2
Total floor area: 13,107.00 m^2

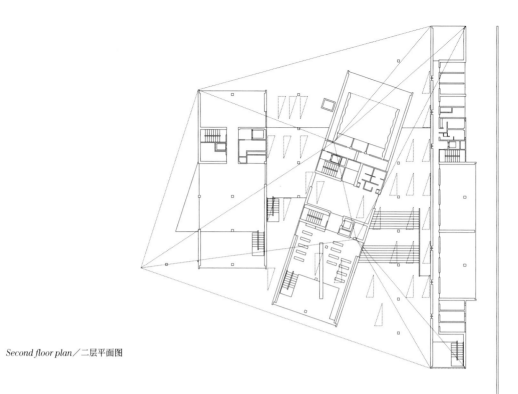

Second floor plan／二层平面图

First floor plan (scale: 1/1000)／一层平面图（比例：1/1,000）

基地位于距离上海不到 200 千米的杭州市郊的良渚文化村，是为了联系当地居民和艺术·文化·教育而建造的综合设施。基地作为 4,000-5,000 年前新石器时代末期长江下游地域存在的古代良渚文化遗址的发现地，久负盛名。良渚文化中形成了拥有城墙的巨大城市。当时农业上已经开始使用犁来耕地，并开始种植稻谷等农作物。不仅是石器，场地上还发现了陶器、玉器、铜器等多种文物。在中国文明文化始祖的探寻上，良渚文化遗址起到了重要的作用。

该建筑的关键中心点是，要在经济成长特别显著的杭州，恢复正在迅速失去的地域社区。我在此尝试了把超越单纯文化设施范畴的"对话的场所"建筑化，形成了象征着家族和地域社区的"大屋顶"之下的空间。大屋顶虽然埋没在森林中，但仍可以作为一个地标来展现建筑的个性。

此外，考虑到西湖和四周包围的运河牵系着杭州人民的心，我将场地附近的河流通过水庭引入建筑，并在水景和河流之间种植了樱花林。到了春天，水边大屋顶的前景就会覆盖着大片盛开的樱花。

建筑根据功能被分成三个体量：美术馆栋、文化设施栋和教育设施栋，排列布置在大屋顶下。另外，通过将中央的文化设施栋偏移进行配置，让三个体块之间留白的檐下空间与建筑物前景中静谧水景的关系，变得充满多样性。大屋顶之下的空间，成为了引发话剧、音乐会、时装秀等人们活动的外部空间。我的目标是最大化地利用基地，实现"与河岸风景的对话"，创造人与文化相接触的良好环境。

安藤忠雄 / 文

pp. 272-273: Elevation view from the South. p. 274: Hand drawing by Tadao Ando. p. 275: The surface of the water by the outside corridor reflects the cherry blossoms in full bloom around it. Image by XU Shiming. p. 277: Aerial view of Liangzhu Art Center. Opposite: Spiral staircase connecting the space above ground and below ground. This page, above: Multi-functional space under big roof. This page, below: Bookshelf that extends to the ceiling. All images on pp. 272–279 by TIAN Fangfang except the specified.

第 272-273 页：从南侧看向建筑。第 274 页，安藤忠雄手绘图。第 275 页，外廊边的水面映出周围盛开的樱花。第 277 页：鸟瞰图。本页：连接地上和地下空间的螺旋楼梯。对页，上：大屋顶下的多功能空间；对页，下：延伸至天花板的书架。